INTERNATIONAL CENTRE FOR MECHANICAL SCIENCES

COURSES AND LECTURES - No. 30

ROBERT GALLAGER
MASSACHUSETTS INSTITUTE OF TECHNOLOGY, CAMBRIDGE

INFORMATION THEORY AND RELIABLE COMMUNICATION

COURSE HELD AT THE DEPARTMENT
OF AUTOMATION AND INFORMATION
JULY 1970

UDINE 1970

SPRINGER-VERLAG WIEN GMBH

Originally published by Springer-Verlag Wien New York in 1972

ISBN 978-3-211-81145-0 ISBN 978-3-7091-2945-6 (eBook)

DOI 10.1007/978-3-7091-2945-6

PREFACE

The following notes were developed by the author in July 1970 in a course on Information Theory at the "Centro Internazionale di Scienze Meccaniche". Except for section 4, which developes a new theory of random trees, most of the material is to be found in expanded form in the author's book, "Information Theory and Reliable Communication", John Wiley and Sons, New York, 1968. A number of the results have been proved here in more satisfying ways which have been developed since the publication of the book. A higher level of mathematical maturity has been assumed here than in the book and an attempt has been made to present some of the deeper aspects of the subject without so much introductory material.

The author would like to express his appreciation to Professor Giuseppe Longo, who organized this set of courses, and to Professor Luigi Sobrero, the Secretary General of C.I.S.M., for their hospitality and for making this work possible.

Finally, he is grateful to the students in the course whose interest and enthusiasm made this a stimulating endeavor.

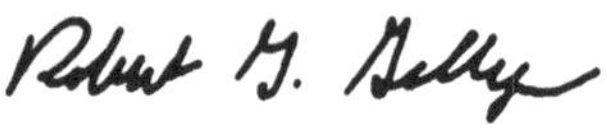

Robert G. Gallager
(Prof. of E.E.,
Mass.Inst. of Tech., U.S.A.)
Udine, Italy, July 10, 1970.

PART I

Introduction

In these notes we shall deal with a number of problems that at first glance seem somewhat disparate. First we shall deal with the noisy channel coding theorem, particularly as related to channels somewhat more complicated than the usual discrete memoryless channel. Then we shall deal with the source coding theorem for sources with a distortion measure defined on them. Finally we shall present some new results on a class of processes called random trees and apply these results to the theory of convolutional codes.

Before going into any of the technical details of these topics, however, it will be appropriate to try to give some engineering perspective into how these topics fit together and how they relate to the problems of communication.

In Fig. 1.1 we show the traditional block diagram of a communication system. The encoder and decoder there are not to be thought of necessarily as the traditional kind of block codes and decoders but rather as any kind of data processing which is included to make the communication more effective. One of the key distinguishing characteristics of information theory which distinguishes it from the older theories of optimal filtering, etc. is the emphasis on the appropriate choice of processing before the channel as well as after the channel.

It is traditional in information theory to break the encoder and decoder each into two pieces, one called the source encoder and decoder and the other called the channel encoder and decoder. The source encoder converts the source output into a sequence of binary digits, the channel encoder then processes these binary digits into a form suitable for transmission over the channel, the channel decoder attempts to recreate the binary data stream that went into the channel encoder, and the source decoder converts these binary digits into a form suitable for the receiver of the communication.

There are obvious merits to making the above split since one can study source coding without worrying about the channel noise and one can study channel coding without worrying about the source. The practical merits of such a split are also obvious and in fact many communication systems are currently being built using the idea of a binary interface between sources and channels. It is less obvious whether there is some irretrievable loss that must be suffered by the separation of source coding and channel coding. One of the principal results of information theory, which will drop out later when we study source coding, is that in a very real sense nothing is lost by separating source coding from channel coding.

Block diagram of a Communication System
Figure 1.1

In the rather copious literature on channel coding and source coding very little is said about how the mathematical models being studied fit into the problems of communication over real channels. Real sources and channels are almost always much more complicated in structure than anything we can deal with analytically. One of the purposes, of course, in studying simple models is to build up a body of techniques that can be used to study progressively more complicated models. A more important purpose, which is often overlooked, is that the simple models might give us enough insight into the interrelations between coding and real channels and sources so that we can design equipment which deals with the main effects and leave the minor effects to be dealt with by engineering perturbations. For this reason, the student of information theory is cautioned that the insight to be gained from a thorough understanding of simple models is far more important than pushing the theory to deal with a slightly more exotic model than has been studied before.

As a final comment in this rather philosophical prelude, it should be pointed out that communication over real channels almost always takes place by the use of analogue waveforms. When we study discrete channels, we are not overlooking this, but simply regarding the encoder as being followed by a digital data modulator which maps a discrete set of samples, once each unit of time, into a discrete set of waveforms. We can improve the system both by improving the encoding and by improving the choice of waveforms. Because this separation into a discrete encoder and a digital data modulator can always be made, it is never really necessary to consider encoders which directly map digital data into waveforms. For some channels, however, it is more convenient to consider waveforms directly. The above point will be clearer by making an analogy to elementary calculus. It is never necessary to use it since discrete approximations can always be used; however such approximations are often terribly inconvenient.

PART II

Measures of Information

For the purposes of studying discrete channel coding, we can view the combination of digital data modulator, channel, and digital data demodulator as simply a discrete channel. In general in information theory, there are no fixed rules on exactly where the channel starts and stops, and therefore it is always appropriate to regard the channel as the part of the system (excluding the source) which we do not care to change at the moment.

The input to the channel at any particular moment is now one of a discrete set of letters which we can view as a sample space. For conceptual purposes we can define a probability for each letter in the sample space. These probabilities might arise from the data and the encoder, or they might be just a useful tool for studying the channel. We shall call such a probability space an ensemble, X. Letting $a_1, a_2, \ldots, a_k$, be the letters in the alphabet, we can denote these probabilities by $P_X(a_i)$, or more compactly by $P(x)$ where x takes on the values $a_1, \ldots, a_K$. Naturally we take the events in this ensemble to be all subsets of the letters, and of course we have $\sum_x P(x) = 1$.

The entropy of an ensemble X is defined to be

$$H(X) = \sum_{x} P(x) \frac{1}{\log P(x)} . \tag{1.1}$$

This can be interpreted as the expected value of the random variable $-\log P(x)$. Entropy has the intuitive interpretation of being a measure of the uncertainty associated with the letter in the ensemble. If the ensemble contains many letters, each of small probability, then $H(X)$ will be large. If there is only one letter with nonzero probability, then $H(X)$ will be zero. In this definition and throughout, we take $0 \log(1/0)$ to be 0.

We can consider the output of the channel at any particular time to form an ensemble Y in the same way. We shall be interested however in the interrelationship between these ensembles, so we consider the product space formed by pairs of input-output letters and let $P_{XY}(a_i, b_\ell)$ be a probability assignment on these pairs where b_ℓ denotes a letter from the ouput. For brevity we denote this probability by $P(x,y)$ but it should be recognized that this leads to mathematical ambiguity if the input and output spaces have common letters. The average mutual information between the X and Y ensemble is then defined as

$$I(X;Y) = \sum_{x,y} P(x,y) \log \frac{P(x|y)}{P(x)} . \tag{1.2}$$

This can trivially be rewritten as

$$I(X;Y) = \sum_{x,y} P(x,y) \log \frac{P(x,y)}{P(x)\,P(y)} \tag{1.3}$$

and from this it is clear that $I(X;Y) = I(Y;X)$.

In terms of the joint X, Y ensemble, we can also define the conditional entropy

$$H(X|Y) = \sum_{x,y} P(x,y) \log \frac{1}{P(x|y)} \tag{1.4}$$

and the joint entropy

$$H(X,Y) = \sum_{x,y} P(x,y) \log \frac{1}{P(x,y)} \tag{1.5}$$

It can be seen that the joint entropy is just the entropy that arises if we regard the joint ensemble as just one bigger ensemble. We can also interpret the conditional entropy as the average uncertainty in X after the output y is known. Finally by breaking up the $\log$ term in (1. 3) into the difference of two terms and simplifying, it is seen that

$$I(X;Y) = H(X) - H(X|Y) \tag{1.6}$$

Thus the average mutual information has the interpretation of being the reduction in uncertainty about x that arises from observing y on the average.

It should be clear that the definitions we have just given apply to arbitrary joint ensembles rather than just the input and output to a channel. We now must make one final definition that applies to three ensembles, say X, Y and Z with a joint probability assignment $P(x, y, z)$. The average conditional mutual information between X and Y conditional on Z is defined as

$$I(X;Y|Z) = \sum_{x,y,z} P(x,y,z) \log \frac{P(x|y,z)}{P(x|z)}. \tag{1.7}$$

As in (1.6), it is easy to see that

$$I(X;Y|Z) = H(X|Z) - H(X|Y,Z). \tag{1.8}$$

Each of these ensembles may be replaced with joint ensembles and there is no limit to the number of relationships that can be found of this type. We will use these relationships as needed in what follows since the reader can easily verify them in the same way as (1.6).

Next we give an extremely important theorem concerning the magnitude of all of these quantities.

Theorem : a) For an ensemble with a sample space of k elements,

$$0 \le H(X) \le \log k \tag{1.9}$$

b) For an arbitrary joint ensemble X, Y,

$$I(X;Y) \ge 0 \tag{1.10}$$

c) For an arbitrary joint ensemble X, Y, Z,

$$I(X;Y|Z) \ge 0 \tag{1.11}$$

In b), there is equality iff X and Y are statistically independent and in c) there is equality iff X and Y conditional on Z are statistically independent (i. e. if $P(x,y|z) = P(x|z)P(y|z)$ for all x, y and all z for which $P(z) > 0$).

Proof : All of these inequalities (except the obvious $0 \leq H(X)$) are proven using the simple inequality

$$\ln w \leq w - 1 \quad \text{for} \quad w > 0$$

with equality iff $w = 1$. We then have

$$H(X) - \log k = \sum_x P(x) \log \frac{1}{kP(x)} \leq$$

$$\leq (\log e)^{-1} \left[\sum_x \frac{1}{k} - \sum_x P(x) \right] = 0 . \qquad (1.12)$$

The base of the logarithm here is arbitrary and simply introduces a multiplicative scale factor on the entropies and mutual informations. The common bases are 2, in which case the unit of information is bits, and e in which case the unit is nats. It can be seen that equality holds in (1.12) iff the elements are equiprobable, with probabilities $1/k$ For part c now, we have

$$-I(X;Y|Z) = \sum_{x,y,z} P(x,y,z) \log \left[\frac{P(x,y|z)}{P(x|y)\,P(y|z)} \right]^{-1} \leq$$

$$\leq (\log e)^{-1} \sum_{x,y,z} \left[\frac{P(x,y,z)P(x|z)\,P(y|z)}{P(x,y|z)} - P(x,y,z) \right] =$$

$$= (\log e)^{-1}\left[\sum_{x,y,z} P(z)P(x|z)P(y|z) - 1\right] = 0.$$

It can be seen that equality holds under the stated conditions. Part b) is the same proof, so it will not be repeated.

As an immediate corollary of part b), we observe from (1.6) that $H(X) \geq H(X|Y)$ and in the same way, from part c) and (1.8) we have $H(X|Z) \geq H(X|Y,Z)$. In other words each additional conditioning done on an ensemble further reduces (or does not increase) its entropy.

As a more important corollary consider the cascaded channel arrangement in fig. 1.2. The variables x, y and z are drawn respectively from ensembles X, Y and Z. The original input ensemble Z is arbitrary. Channel one is defined by a conditional probability assignment $P(x|z)$ and channel two is defined by a conditional probability assignment $P(y|x)$. We assume that the output of channel two is statistically related to z only through the input to the second channel, x. More precisely, we assume that for each z, x of non zero probability and for each y,

$$P(y|x) = P(y|x,z) \tag{1.13}$$

This is equivalent to the condition $P(y,z|x) = P(y|x)P(z|x)$. Thus from part c) of the theorem, reversing the roles of X and Z, we see that $I(Z;Y|X) = 0$.

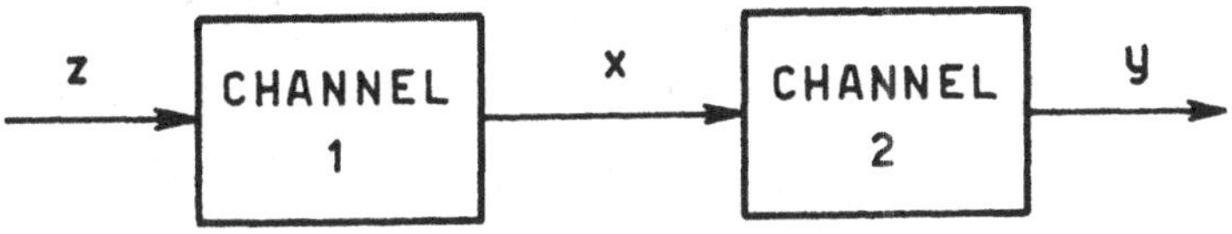

Cascaded Channels
Fig. 1.2

Next we observe that in general

$$I(ZX;Y) = I(Z;Y) + I(X;Y|Z) = I(X;Y) + I(Z;Y|X).$$

Since $I(Z;Y|X) = 0$, this reduces to

$$I(X;Y) = I(Z;Y) + I(X;Y|Z). \tag{1.14}$$

Since all of the above quantities are positive, this yields the following two inequalities :

$$I(X;Y) \geq I(Z;Y) \tag{1.15}$$

$$I(X;Y) \geq I(X;Y|Z). \tag{1.16}$$

Both of these results are important. The first, for reasons that will soon become apparent, is often called the data processing theorem, and the second is a slightly camoflaged way of saying that the mutual information on a channel is a convex function of the input probabilities. First we discuss (1.15). This states that the mutual information across a cascade of channels can never be greater than the mutual information on the second channel. By interchanging the role of Z and Y in (1.13), it is seen that $I(Z;X) \geq I(Z;Y)$ also, from which it immediately follows that the mutual information over a cascade of channels can not exceed the mutual information over any channel in the cascade. To put it another way, the mutual information about an input to a cascade of channels decreases as we move down the line. This result becomes more surprising if we view the encoder and decoder in Fig. 1.1 as channels, for then we see that the mutual information between source and destination can not exceed the mutual information over the channel. In particular the information about the source after decoding can not exceed the mutual information before decoding. Decoders make decisions and put the data into a useful form, but they cannot create new information about the source output. Thinking about this is very helpful in improving ones intuitive understanding of the nature of this information measure. The importance of this result seems to have first been appreciated by Woodward who used it as the foundation of his theory of sufficient receivers.

To interpret (1.16), recall that channel two was defined in terms of the conditional probability assignment $P(y|x)$.

The mutual information on channel two is a function both of $P(y|x)$ and of the probability assignment on X, which we now denote by $Q(x)$. We can think of this probability assignment as a vector, $\vec{Q} = (Q(a_1), \ldots, Q(a_K))$ and we can think of the probability assignment $P(y|x)$ as another vector (of higher dimensionality). Thus, $I(X;Y)$ is in reality a function of the vector $\vec{Q}$ and the vector $\vec{P}$, so we can write it as

$$I(X;Y) = f(\vec{Q}, \vec{P}).$$

Now let Z be a binary ensemble, taking on the value 0 with probability λ and the value 1 with probability $1 - \lambda$. Let $Q_0(x)$ and $Q_1(x)$ denote the conditional probabilities of x conditional on $z = 0$ and $z = 1$ respectively. Then $\vec{Q} = \lambda \vec{Q}_0 + (1-\lambda)\vec{Q}_1$. Finally with a little rearrangement of (1.7), we see that

$$I(X;Y|Z) = \lambda f(\vec{Q}_0, \vec{P}) + (1-\lambda) f(\vec{Q}_1, \vec{P}) \tag{1.17}$$

Thus (1.16) takes the form

$$f(\lambda \vec{Q}_0 + (1-\lambda)\vec{Q}_1, \vec{P}) \geq \lambda f(\vec{Q}_0, \vec{P}) + (1-\lambda) f(\vec{Q}_1, \vec{P}) \tag{1.18}$$

and we see that the mutual information on a channel is a convex $\cap$ function of the input probability assignment. A similar argument, using Z as a binary indicator of two different conditional probability assignments, shows us that $f(\vec{Q}, \vec{P})$ is a convex $\cup$ function of $\vec{P}$. We use convex $\cap$ and convex $\cup$ to denote what is usually denoted by concave and convex since nearly everyone gets confused on the distinction between convex and concave.

Next we shall use these convexity results to discuss how to find the capacity of a channel. We assume as before that the channel is described by a set of transition probabilities $P(y|x)$ and that $\vec{Q} = (Q(a_1),\ldots,Q(a_K))$ is an input probability assignment for the channel. We shall discuss more fully later the sense in which a transition probability assignment defines a channel, but for the moment it is only necessary to understand that the input assignment $\vec{Q}$ is not a part of the channel but only has to do with how the channel is used. There is thus some rationale for defining the capacity of the channel as the maximum average mutual information that can be transmitted over it, with the maximization taken over all input probability assignments. This definition is of course given significance later by the noisy channel coding theorem. We then have capacity C defined by

$$C = \max_{\vec{Q}} f(\vec{Q},\vec{P}) \tag{1.19}$$

where the maximum is over all probability vectors $\vec{Q}$. The maximum must exist since the function is continuous and the region over which the maximization is performed is closed and bounded. The partial derivatives of f with respect to the components of $\vec{Q}$ are thus given by

$$\frac{\partial f(\vec{Q},\vec{P})}{\partial Q(a_i)} = \frac{\partial}{\partial Q(a_i)} \sum_{y,x} Q(x) P(y) \log \frac{P(y|x)}{\sum_{x'} P(y|x')} = f_i(\vec{Q},\vec{P}) - \log e$$

where

$$f_i(\vec{Q},\vec{P}) = \sum_y P(y|a_i) \log \frac{P(y|a_i)}{\sum_x Q(x) P(y|x)} . \qquad (1.20)$$

The quantities f_i defined above can be interpreted as the mutual information random variable averaged over the Y space but not the X space. Considering f as a function of $\vec{Q}$ alone now, we can write the equation of a hyperplane which is tangent to f at the point $\vec{Q}$. As a function of some other point $\vec{Q}_1$, it is

$$f(\vec{Q},\vec{P}) + \sum_{i=1}^{k} \left[Q_1(a_i) - Q(a_i)\right] \left[f_i(\vec{Q},\vec{P}) - \log e\right]. \qquad (1.21)$$

Because of the convexity of f over the region where $\vec{Q}$ is a probability vector we know that the function lies below the hyperplane in this region. Thus we have

$$f(\vec{Q}_1,\vec{P}) \le f(\vec{Q},\vec{P}) + \sum_i \left[Q_1(a_i) - Q(a_i)\right] \left[f_i(\vec{Q},\vec{P}) - \log e\right]. \qquad (1.22)$$

We can simplify this equation now by observing that the sum over the $\log e$ term is zero since it is multiplied by the difference of two probability assignments. Also, we have

$$f(\vec{Q},\vec{P}) = \sum_{i=1}^{k} Q(a_i) f_i(\vec{Q},\vec{P}). \qquad (1.23)$$

Substituting this in (1.22), we obtain

$$f(\vec{Q}_1, \vec{P}) \leq \sum_i Q_1(a_i)\, f_i(\vec{Q}, \vec{P}) \leq \max_i f_i(\vec{Q}, \vec{P})$$

The second bound above is uniformly valid for all $\vec{Q}_1$ and is thus also a bound on C, so that

$$(1.25) \qquad f(\vec{Q}, \vec{P}) \leq C \leq \max_i f_i(\vec{Q}, \vec{P}).$$

<u>Theorem</u> : Necessary and sufficient conditions on $\vec{Q}$ to achieve capacity are that for some constant C,

$$(1.26) \qquad f_i(\vec{Q}, \vec{P}) \leq C \quad \text{for all } i, \quad 1 \leq i \leq k$$

with equality for all i such that $Q(a_i) > 0$. Furthermore the constant C above is the channel capacity.

<u>Proof</u> : First suppose that a probability vector $\vec{Q}$ satisfies (1.26). Then, from (1.24) we see that

$$(1.27) \qquad f(\vec{Q}, \vec{P}) = \max_i f_i(\vec{Q}, \vec{P}) = C$$

where C is the constant in (1.26). However (1.25) asserts that this C is the channel capacity, so we have established the sufficiency of the condition. For the necessity, consider any $\vec{Q}$ that does not satisfy (1.26) for any constant C. Then there are two input letters, say a_i and a_j for which $f_i(\vec{Q}, \vec{P}) < f_j(\vec{Q}, \vec{P})$ and such that $Q(a_i) > 0$. We then see from the equation of the tangent hyperplane in (1.21) that a slight decrease in $Q(a_i)$

with a corresponding increase in $Q(a_j)$ will yield a $\vec{Q}_1$ with larger F than Q, thus completing the proof.

For a channel with sufficient symmetry one can often guess the input probabilities that yield capacity and then check the result from the theorem. For more complicated channels the easiest way to find capacity is usually a computer hill climbing technique. One starts with an arbitrary $\vec{Q}$ and then successively increases those probabilities for which F_i is large and decreases those for which it is small. Sometimes one or more inputs are so bad that their probability gets reduced to 0 without bringing them up to the level of the others, which leads to the inequality condition in (1.27). For a more complete discussion of ways to find capacity, see Gallager, Information Theory and Reliable Communication, chapter 4, and for generalizations of these maximization techniques, see any text on convex programming.

Finite State Channels

So far we have been somewhat vague about how to describe a channel mathematically. In this section we shall clarify this matter and develop some results about finite state channels. Unfortunately, relatively little is known about such channels except for some very general results that are rather hard to apply so we shall raise more questions than we shall answer. We shall restrict ourselves here to discrete channels to avoid fussing about numerous mathematical details that only obscure one's vision.

The input to a discrete channel is a sequence of letters, denoted by $\ldots x_{-1}, x_0, x_1, \ldots$. These letters are drawn from a common alphabet, say the set of integers $(0, 1, \ldots, k-1)$. Similarly the output is a sequence of letters, denoted by $\ldots, y_{-1}, y_0, y_1, \ldots$. These are drawn from the alphabet $(0, 1, \ldots, J-1)$. The channel is defined by specifying a probability measure on the output sequence for each choice of input sequence. We shall assume that the channel is non-anticipatory which means that for each choice of past and present inputs and each choice of past outputs a conditional probability assignment on the present output can be made. For a given time n we can summarize all of the relevant past into a state variable s_{n-1} and then the output from the channel y_n and the new state s_n can be defined by a probability measure on s_n and y_n conditional on x_n and s_{n-1}. Such a model is perfectly general, but we now make some further restrictions. First we assume that the state space is finite with A elements, denoted by $(0, 1, \ldots, A-1)$. Thus we can describe the channel by a conditional probability assignment $P(y_n, s_n | x_n, s_{n-1})$. Next we assume that the channel is stationary so that P, as a function of four discrete variables, does not depend on the time n. When we say that the channel is described by P we mean that, conditional on x_n and s_{n-1} the pair y_n, s_n is statistically independent of all other past inputs, outputs, and states. Physically the states can be considered as a quantized version of the variables that introduce memory of the past into the channel. Such variables are the fading level of a fading channel and the effects of filtering on the previous inputs, for example. Unfortunately, it turns out that for some channels, even a very fine quantization of these variables markedly changes the information theoretic behavior of the channel. A discrete memoryless channel is defined as a finite state channel with only one element in the state space. Thus, conditional on x_n, y_n is independent of the past.

There are several different formulations for a finite state channel which are all equivalent. In order to show this, we first rewrite the probability assignment as

$$(2.1)\qquad P(y_n, s_n | x_n, s_{n-1}) = P(s_n | x_n, s_{n-1}) P(y_n | x_n, s_{n-1}, s_n)$$

In this formulation, then, each transition from one state to another has a set of transition probabilities from channel input to output associated with it. To get a different formulation, we can define a new state space S' as the direct product of the old state space with itself. The state at time n in the new system, s'_n is then the pair of states s_n and s_{n-1}. The system is then described by $P(s'_n | x_n, s'_{n-1}) P(y_n | x_n, s'_n)$. In other words, we now have a set of transition probabilities associated with each terminal state rather than with each transition between states.

As yet another formulation, let s''_n be the pair s_n, y_n. Then y_n becomes a deterministic function of s''_n and the channel is described by this function and by $P(s''_n | x_n, s''_{n-1})$. We have now shown that these new formulations, by a redefinition of the state space, describe any channel covered by the original definition. But without changing the state space, the new formulations are clearly special cases of the old so in fact all three formulations are equivalent. The last formulation is the one used by Blackwell, Breiman, and Thomasian, but the first two are preferred here because they are more natural physically.

We say that a channel has no intersymbol interference memory if the state at time n, conditional on the state at time $n-1$ is independent of x_n(i.e. if $P(s_n \mid x_n, s_{n-1}) = P(s_n \mid s_{n-1})$). In this case, the state sequence is described as a homogeneous Markov chain and the structure of the channel is conceptually fairly simple. This special case only appears if one uses either of the first two formulations above. It cannot be meaningfully described in terms of the Blackwell formulation. Fading channels and digital channels with burst errors can be meaningfully described by models of this type.

We say that a channel has only intersymbol interference memory if the state at time n is a deterministic function of x_n and s_{n-1}. Such a model can be used for telephone lines where the major effect of memory is due to the filtering in the channel and each input becomes spread out in time. We shall also refer to these channels as channels where the state is specified at the transmitter.

A slightly more general special case is that where the state is specified by the transmitter and receiver together, i.e. where s_n is a deterministic function of x_n, y_n, and s_{n-1}. Analogous to the channel with state specified at the transmitter is a channel with state specified at the receiver, i.e. where s_n is a deterministic function of s_{n-1} and y_n.

Some insight into the problems of defining capacity for a finite state channel can be seen from the example below of an intersymbol interference only channel. As indicated in the figure if the channel is in state zero, it remains there with an input of 0 or 1 and goes to state 1 with an input of 2. Once the channel is in state 1, it always remains there. There is an analogy to a childs toy here, where inputs 0 and 1 represent playing with the toy and input 2 represents breaking the toy. There is a great deal of immediate fun (information) to be gained from breaking the toy, but that destroys all future fun (information). If the channel starts in state 0, then one bit per use can be transmitted indefinitely by using only inputs 0 and 1. More information can be transmitted at any given time by using input two, but on a long term basis, this is unprofitable. If the channel starts in state 1, then clearly nothing can be transmitted.

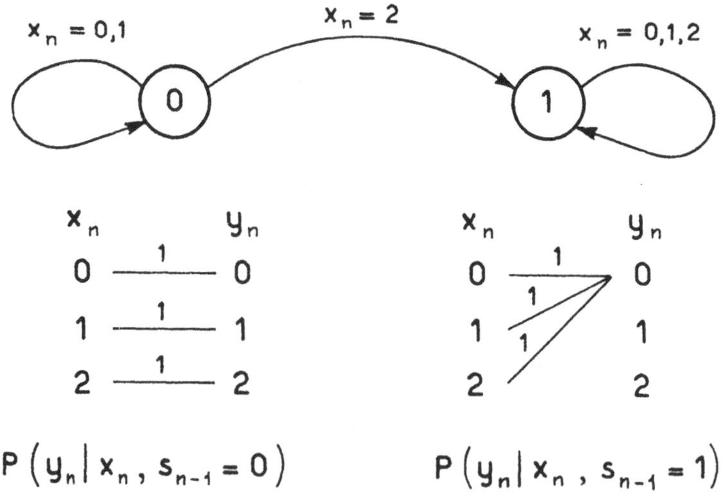

Child's Toy Channel
Figure 2.1

If we now let $\vec{x}$ denote the sequence $x_1, x_2, \ldots, x_N$ and $\vec{y}$ denote the sequence $y_1, y_2, \ldots, y_N$, then we can inductively define the probability of an output sequence conditional on an input sequence and an initial state by

$$P_N(\vec{y}, s_N | \vec{x}, s_0) = \sum_{s_{N-1}} P(y_N, s_N | x_N, s_{N-1}) P_{N-1}(\vec{y}, s_{N-1} | \vec{x}, s_0). \quad (2.2)$$

In these expressions the sequence lengths for $\vec{x}$ and $\vec{y}$ are indicated by the subscript on P. We then have

$$P_N(\vec{y} | \vec{x}, s_0) = \sum_{s_N} P_N(\vec{y}, s_N | \vec{x}, s_0) \quad (2.3)$$

Note however that except in the case of a channel without intersymbol interference, it is impossible to define a probability of an output sequence conditional only on the input sequence in terms of the channel alone. The difficulty is that the distribution of the initial state, s_0 will depend on the input distribution which is not a part of the channel definition.

We now define an upper and lower capacity for the channel at each different blocklength. The upper capacity is given by

$$\bar{C}_N = \frac{1}{N} \max_{Q_N(\vec{x})} \max_{s_0} f(\vec{Q}_N, \vec{P}_N(s_0)) \quad (2.4)$$

where

$$f(\vec{Q}_N, \vec{P}_N(s_0)) = \sum_{\vec{x}} \sum_{\vec{y}} Q_N(\vec{x}) P_N(\vec{y} | \vec{x}, s_0) \log \frac{P_N(\vec{y} | \vec{x}, s_0)}{\sum_{\vec{x}'} Q_N(\vec{x}') P_N(\vec{y} | \vec{x}', s_0)}. \quad (2.5)$$

The lower capacity is similarly given by

$$\underline{C}_N = \frac{1}{N} \max_{Q_N(\bar{x})} \min_{s_0} f\left(\vec{Q}_N, \vec{P}_N(s_0)\right) \tag{2.6}$$

We then define the limiting upper and lower capacities of the channel by

$$\bar{C} = \lim_N \bar{C}_N \quad ; \quad \underline{C} = \lim_N \underline{C}_N \tag{2.7}$$

The existence of these limits and bounds on their value are established by Theorem 4.6.1 in Gallager, Information Theory and Reliable Communication, which states that

$$\bar{C} = \inf\left(\bar{C}_N + (\log A)/N\right) \tag{2.8}$$

$$\underline{C} = \sup\left(\underline{C}_N - (\log A)/N\right). \tag{2.9}$$

From this theorem and the fact that $\bar{C} \geq \underline{C}$, we get the relation that for any N,

$$\bar{C}_N + (\log A)/N \geq \bar{C} \geq \underline{C} \geq \underline{C}_N - (\log A)/N. \tag{2.10}$$

We shall not prove the above theorem here since it is just a matter of some rather messy juggling of information equalities and inequalities. The major point of the proof is to split up the mutual information for some long block length N into the mutual information for the first n digits of the length and the last $N-n$ Due to the memory of the channel, there is also some cross-coupling between these shorter lengths, but since there are only A possible states, this cross-coupling term is at most $\log A$, giving rise to that term in the result.

There is little justification for arguing that either $\bar{C}$ or $\underline{C}$ is <u>the</u> capacity of the channel. $\bar{C}$ is the limiting mutual information per digit that could be transmitted if the transmitter had its choice of the initial state for the channel to be in. $\underline{C}$ is the limiting mutual information that could be transmitted per digit if the transmitter first chose its input distribution and then the channel was put into the worst possible initial state for that distribution. In most of the literature, $\underline{C}$ is taken to be the capacity of the channel, but this is completely meaningful only when $\underline{C} = \bar{C}$.

Unfortunately no necessary and sufficient conditions are known under which $\underline{C} = \bar{C}$. There is, however, a rather broad class of channels, known as indecomposable channels, for which $\underline{C}$ is always equal to $\bar{C}$. In order to define these, we first define the probability of a state s_N at time N conditional on the state s_0 at time 0 and on the input $\vec{x} = x_1, x_2, \ldots, x_N$ in the interim,

$$q\left(s_N \mid \vec{x}, s_0\right) = \sum_{\bar{y}} P_N\left(\bar{y}, s_N \mid \vec{x}, s_0\right).$$

A finite state channel is defined to be indecomposable if for each $\epsilon > 0$, there exists an N_0 such that for all $N \geq N_0$,

$$\left| q\left(s_N \mid \vec{x}, s_0\right) - q\left(s_N \mid \bar{x}, s_0'\right) \right| \leq \varepsilon \qquad (2.11)$$

for all $\vec{x}$, s_N, s_0, s_0'. Thus a channel is indecomposable if its memory of the initial state dies out with increasing time. The above definition is intuitively pleasing, but is somewhat difficult to use in determining whether or not a channel is indecomposable. To obtain a simpler criterion to work with, we first suppress the $\vec{x}$ variable above, remembering that the subsequent argument is all to be regarded as conditional on a fixed input sequence. We then define the distance between two initial states, measured at some later time N, to be

$$d_N(s_0', s_0'') = \sum_{s_N} \left| q(s_N | s_0') - q(s_N | s_0'') \right| . \tag{2.12}$$

If the initial dependence on s_0 is going to decrease with N, then this quantity should go to zero with increasing N.

Lemma : Suppose that for some $n > 0$ and for some $\delta > 0$, and for each $N \geq 0$, there is some choice of s_{N+n} such that

$$q(s_{N+n} | s_N) \geq \delta \quad \text{for all values of } s_N \tag{2.13}$$

Then

$$d_N(s_0', s_0'') \leq 2(1-\delta)^{(N/n)-1} \tag{2.14}$$

Proof :

$$d_{N+n}(s_0', s_0'') = \sum_{s_{N+n}} \left| \sum_{s_N} q(s_{N+n} | s_N)\left[q(s_N \; s_0') - q(s_N \; s_0'')\right] \right| . \tag{2.15}$$

Define

$$a(s_{N+n}) = \min_{s_N} q(s_{N+n} | s_N). \tag{2.16}$$

Observing that

$$\sum_{s_N} a(s_{N+n}) \left[q(s_N | s_0') - q(s_N | s_0'') \right] = 0 \tag{2.17}$$

we can rewrite (2.15) as

$$d_{N+n}(s_0', s_0'') = \sum_{s_{N+n}} \left| \sum_{s_n} \left[q(s_{N+n} | s_N) - a(s_{N+n}) \right] \left[q(s_N | s_0') - q(s_N | s_0'') \right] \right| \tag{2.18}$$

$$\leq \sum_{s_{N+n}} \sum_{s_n} \left[q(s_{N+n} | s_N) - a(s_{N+n}) \right] \left| q(s_N | s_0') - q(s_N | s_0'') \right| \tag{2.19}$$

$$= \left[1 - \sum_{s_{N+n}} a(s_{N+n}) \right] d_N(s_0', s_0'') \tag{2.20}$$

$$\leq (1 - \delta)\, d_N(s_0', s_0''). \tag{2.21}$$

In (2.19), we have upper bounded the magnitude of a sum by the sum of the magnitudes and recognized that the first term in brackets is always positive. In (2.21) we have used the fact that $a(s_{N+n})$ is at least δ for some s_{N+n}. We can now iterate this result to see that

$$d_{kn}(s_0', s_0'') \leq (1 - \delta)^k d_0(s_0', s_0'').$$

Recognizing finally that $d_0 = 2$ and that (2.20), used for arbitrary n asserts that d_N is nonincreasing in N, we have the final result.

We now have the following theorem :

Theorem : A necessary and sufficient condition for a finite state channel to be indecomposable is that for some fixed n and each $\vec{x}$, there exists a $\delta > 0$ such that

$$q(s_n | \vec{x}, s_0) > \delta \tag{2.22}$$

for all s_0, all $\vec{x}$, and some s_n depending on $\vec{x}$.

Proof: The sufficiency of the condition follows immediately from the lemma. For the necessity, pick in (2.11) to be less than $1/A$ and pick n larger than the N_0 of (2.11). For a given s_0 and $\vec{x}$, pick s_n so that $q(s_n | \vec{x}, s_0) \geq 1/A$. Then from (2.11), $q(s_n | \vec{x}, s_0') \geq$ $\geq 1/2A$ for all s_0', completing the proof.

In applying this theorem, one wants to know how large must be before one can stop searching for an n that satisfies the conditions of the theorem. Thomasian has shown that if such an n exists, then such an $n < 2^{(A^2)}$ must exist. To derive this result, define

$$T_{n,\vec{x}}(s_0, s_n) = \begin{cases} 1 ; & q(s_n | \vec{x}, s_0) > 0 \\ 0 ; & \text{otherwise} \end{cases} \tag{2.23}$$

The range of this function, for given n and $\vec{x}$, is a set of A^2 points, each of which is mapped into 1 or 0. Thus, this function must be one of $2^{(A^2)}$ different choices. Thus, as a sequence of functions in n for a fixed $\vec{x}$, the sequence must either terminate (it terminates if there is an s_n for which the condition of the theorem is satisfied for that $\vec{x}$), or the sequence has two functions the same, say T_i and T_n for $i < n < 2^{(A^2)}$. If such is the case, then choosing $x_{n+k} = x_{i+k}$ for all positive integers k will assure that $T_{n+k} = T_{i+k}$. In this case, we have exhibited an $\vec{x}$ which violates the conditions of the theorem for all n .

Even with this result, one must test an extraordinary number of input sequences. To streamline the test somewhat, let $\vec{x}_1$ be the first i digits of a sequence $\vec{x}$ of length n and let $\vec{x}_2$ be the final $n-i$ digits. Then it is easy to verify that

$$T_{n,\vec{x}}(s_0, s_n) = \begin{cases} 1 & \text{if } T_{i,\vec{x}_1}(s_0, s_i)\, T_{n-i,\vec{x}_2}(s_i, s_n) = 1 \text{ for some } s_i \\ 0 & \text{otherwise} \end{cases} \tag{2.24}$$

From the previous test, we see that the channel will be indecomposable unless there is some n , $\vec{x}$, and i for which the function $T_{n,\vec{x}}$ is the same as the function $T_{i,\vec{x}_1}$. Thus, one may simply calculate all possible functions of this type that can arise from the given channel and see if the combination of two of them (for which the function is not 1 for all values of the first argument and some value of the second argument) as in (2.24) gives rise to the first function again. If this happens, the channel is not indecomposable.

Theorem : If a channel is indecomposable, then $\underline{C} = \bar{C}$. The proof of this theorem can be found in Gallager p. 109 and therefore we shall not repeat it here. The idea is simple although the manipulations are somewhat involved. The point is that the effect of the initial state dies out with time and thus only effects the information transfer over a short period of time. Since C is defined as the limit over an arbitrarily large time, the initial transient effect of the initial state vanishes over the long interval.

The previous theorem says nothing about indecomposability being necessary to have $\underline{C} = \overline{C}$ and in fact there are many decomposable channels for which $\underline{C} = \overline{C}$. A particularly interesting example of such channels forms a subclass of the channels with intersymbol interference memory only. Suppose that there exists an input sequence which will drive the channel into a known state and suppose also that any state can be reached from any other state by an appropriate input sequence. Then, one can use the appropriate input sequence to drive the channel into the best possible state and from that point transmit at $\overline{C}$. Over a sufficiently long interval, the loss of information in the short interval required to drive the channel into the best state is negligible, and thus $\underline{C} = \overline{C}$.

Finally, for a channel with no intersymbol interference memory, the test for indecomposability becomes quite simple, since it is independent of the input sequence. Thus, such a channel is indecomposable iff the Markov chain describing the state sequence is ergodic.

We will now proceed to prove a coding theorem for finite state channels. The major part of this result is the same whether we are dealing with finite state channels, discrete memoryless channels, or channels with arbitrary input and output spaces. The only special things about finite state channels involve first finding a way to handle the problem of an unknown initial state, and second finding a way to interpret the result that we obtain.

We shall handle the first problem, that of the initial states, by an artifice. We shall construct an encoder and a decoder under the assumption that there is a probability distribution on the initial states, the assumption being that the initial states are equally likely. Then, we shall find an upper bound on the probability of error that can be obtained, under this assumption of equally likely states. This probability of error will be $1/A$ times the sum of the error probabilities for the given code and decoder conditional on each of the possible initial states. Thus, conditional on any given initial state, the error probability can be at most A times the error probability averaged over the initial states. We thus define $P_N(\vec{y}\,|\,\vec{x})$ as

$$P_N(\vec{y}\,|\,\vec{x}) = \frac{1}{A} \sum_{s_0} P_N(\vec{y}\,|\,\vec{x}, s_0) \tag{2.25}$$

We will now find an upper bound to the probability of error that can be achieved on this channel for a code consisting of M code words. The method of doing this will be to consider an ensemble of codes, each with its own decoder, and to evaluate the average probability of error over this ensemble of codes and decoders. The rule for decoding for each code will be maximum likelihood. That is, if $\vec{x}_1, \vec{x}_2, \ldots, \vec{x}_M$ denote the code words in a code, then the decoder will take the received sequence $\vec{y}$ and evaluate $P_N(\vec{y}\,|\,\vec{x}_m)$ for each m, $1 \le m \le M$. It will choose for the decoded message that m for which $P_N(\vec{y}\,|\,\vec{x}_m)$ is maximum. Since we are interested in upper bounding the probability of error, we assume that an error always occurs if the two largest terms are equal. The ensemble of codes is generated by choosing each code word independently according to some fixed probability assignment, $Q_N(\vec{x})$. We thus have a probability system in which the M code words are chosen according to the distribution above, a message m is selected for transmission, the randomly selected code word $\vec{x}_m$ is transmitted, and $\vec{y}$ is received according to the probability $P_N(\vec{y}\,|\,\vec{x}_m)$. A decoding error occurs if $P_N(y\,|\,x_{m'}) \ge P_n(\vec{y}\,|\,\vec{x}_m)$ for one or more values of m'. It will not be necessary to specify a probability distribution on the message set since we shall evaluate the average probability of error conditional on a particular message, say m, being transmitted. Denoting this average error probability by $\bar{P}_{e,m}$ we have

$$\bar{P}_{e,m} = \sum_{\vec{x}_m} \sum_{\vec{y}} Q_N(\vec{x}_m)\, P_n(\vec{y}\,|\,\vec{x}_m)\, Pr(\text{error}\,|\,m, \vec{x}_m, \vec{y}) \tag{2.26}$$

The final term in this equation is the average error

probability conditional on message m entering the encoder, cord word $\vec{x}_m$ being chosen for the mth code word and $\vec{y}$ being the received sequence. The fact that the entire set of code words must be selected and the receiver informed as to their choice before any communication can begin does not make it any the less meaningful to define the quantities above, since we have a perfectly well defined probability space and such conditional probabilities are well defined. Now let $A_{m'}$ be the event that $\vec{x}_{m'}$ is selected in such a way that $P_N(\vec{y}\,|\,\vec{x}_{m'}) \geq$ $\geq P_N(\vec{y}\,|\,\vec{x}_m)$. We are considering these events in the space conditional on m, $\vec{x}_m$, and $\vec{y}$. Thus we see that

$$Pr(\text{error}\,|\,m, \vec{x}_m, \vec{y}) = Pr\left(\bigcup_{m' \neq m} A_{m'}\right) \qquad (2.27)$$

Clearly the above probability can be bounded both by the sum of the probabilities of the individual events and by 1. Unfortunately neither bound is adequate in general. Upper bounding by the sum of the probabilities will yield a result which is very much greater than 1 when the noise is very bad. These terms will then contribute too much to the average, and will in fact dominate the whole behavior of the resulting bound. Similarly a bound of 1, when substituted back into (2.26)will reduce the entire bound on error probability to 1, which is a rather useless result to work so hard for. Thus we take a compromise between these alternatives and use

$$\Pr\left(\text{error} \mid m, \vec{x}_m, \vec{y}\right) \le \left[\sum_{m' \ne m} \Pr(A_{m'})\right]^{\varrho}; \quad 0 < \varrho \le 1. \tag{2.28}$$

We observe that when the union bound is greater than 1, this bound is also greater than 1 and thus valid, and when the union bound is less than 1, this bound exceeds the union bound, and is thus valid again, so it is valid in all cases. It appears that it is a very crude bound, but it is not so bad as it appears. Most of the contribution to error probability comes from those cases where $\Pr\left(\text{error} \mid m, \vec{x}_m, \vec{y}\right)$ is close to 1, and here the bound is quite tight. The bound simply depresses the effect of very bad noise sequences at the expense of very good noise sequences, and neither are very important.

Since the event $A_{m'}$ depends solely on the choice of $\vec{x}_{m'}$ conditional on the given m, x_m, and y, we can upper bound its probability quite easily. We have

$$\Pr(A_{m'}) = \sum_{\vec{x}_{m'} \,:\, P_N(\vec{y} \mid \vec{x}_{m'}) \ge P_N(\vec{y} \mid \vec{x}_m)} Q_N(\vec{x}_{m'})$$

$$\le \sum_{\vec{x}_{m'}} Q_N(x_{m'}) \left[\frac{P_N(\vec{y} \mid \vec{x}_{m'})}{P_N(\vec{y} \mid \vec{x}_m)}\right]^s; \quad 0 < s \le 1. \tag{2.29}$$

The above bound is derived by noting that the ratio above is greater than or equal to 1 for each term in the original sum. We then sum over all $\vec{x}_{m'}$, thus further increasing the right hand side. This bound is again tighter than it appears to be, since the important contributions to error probability come when $P_N(\vec{y} \mid \vec{x}_{m'})$ and $P_N(\vec{y} \mid \vec{x}_m)$ are almost equal. Thus again, we are merely finding a convenient way to avoid fussing about the terms that are not important anyway.

Combining (2. 26), (2. 28), and (2. 29), we obtain

$$\bar{P}_{e,m} \leq \sum_{\vec{y}} \sum_{\vec{x}_m} Q_N(\vec{x}_m) P_N(\vec{y} \mid \vec{x}_m) \left[\sum_{m' \neq m} \sum_{\vec{x}_{m'}} Q_N(\vec{x}_{m'}) \frac{P_N(\vec{y} \mid \vec{x}_{m'})^s}{P_N(\vec{y} \mid \vec{x}_m)^s} \right]^{\rho} \quad (2.30)$$

At this point we observe that $\vec{x}_{m'}$ is just a variable of summation, and thus the sum over m' is just a sum over $M-1$ equal terms. Thus we have

$$\bar{P}_{e,m} \leq \sum_{\vec{y}} \left(\sum_{\vec{x}_m} Q_N(\vec{x}_m) P_N(\vec{y} \mid \vec{x}_m)^{1-s\rho} \right) (M-1)^{\rho} \left[\sum_{\vec{x}} Q_N(\vec{x}) P_N(\vec{y} \mid \vec{x})^s \right]^{\rho} . \quad (2.31)$$

This bound is valid for all s, $0 < s \leq 1$, and all ρ, $0 < \rho \leq 1$, so we obtain the tightest bound by optimizing over these parameters. The optimum value of s is at $s = 1/(1+\rho)$. This choice has the added advantage of making the sum over $\vec{x}_m$ the same as the sum over $\vec{x}$, so we have

$$\bar{P}_{e,m} \leq (M-1)^{\rho} \sum_{\vec{y}} \left[\sum_{\vec{x}} Q_N(\vec{x}) P_N(\vec{y} \mid \vec{x})^{\frac{1}{1+\rho}} \right]^{1+\rho} . \quad (2.32)$$

This bound is valid for all channels for which $P_N(\vec{y} \mid \vec{x})$ can be defined. Also, as expected, it does not depend on which m enters the encoder. Thus, if we define $\bar{P}_e = (1/M) \sum_m \bar{P}_{e,m}$, we see that $\bar{P}_e$ is also bounded by (2. 32). Finally there must exist at least one code in the ensemble for which P_e is bounded by (2. 32).

It is also convenient to have a bound on error probability that applies uniformly to each code word. To accomplish this, start with a code of $2M$ code words for which (2. 32) upper bounds the error probability (with $2M$ substituted for M). Then, delete the M code words of highest error probability. This deletion cannot increase the error probability of any of the remaining code words, and all the remaining code words must have error probabilities no more than twice the average. Thus, we have a code with M code words for which, for each m,

$$P_{e,m} \le 4 M^{\varrho} \sum_{\vec{y}} \left[\sum_{\vec{x}} Q_N(\vec{x}) P_N(\vec{y} \mid \vec{x})^{\frac{1}{1+\varrho}} \right]^{1+\varrho} . \tag{2. 33}$$

Now let us reinsert (2. 25) for $P_N(\vec{y} \mid \vec{x})$. The bound is difficult to interpret and work with with this sum on s_0 in the middle of the bound. We can get the sum on the outside, however, by applying the two elementary inequalities

$$\left(\sum_i a_i\right)^r \le \sum_i a_i^r \ ; \quad a_i \ge 0 \ , \ r \in (0,1] \tag{2. 34}$$

$$\left(\sum_{i=1}^{A} \frac{1}{A} a_i\right)^r \le \sum_i \frac{1}{A} a_i^r \ ; \quad a_i \ge 0 \ , \ r \ge 1. \tag{2. 35}$$

Applying these, we obtain

$$P_{e,m} \le A^{\varrho-1} 4 M^{\varrho} \sum_{s_0} \sum_{\vec{y}} \left[\sum_{\vec{x}} Q_N(\vec{x}) P_N(\vec{y} \mid \vec{x}, s_0)^{\frac{1}{1+\rho}} \right]^{1+\rho} . \tag{2. 36}$$

Finally, we upper bound the sum on s_0 by A times the maximum term and we observe that the error probability conditional on a given intial state can be no more than A times the average over the initial state. Thus, for each code word and each initial state, we have

$$P_{e,m}(s_0) \leq A^{1+\varrho}\, 4\, M^{\varrho} \max_{s_0} \sum_{\vec{y}} \left[\sum_{\vec{x}} Q_N(\vec{x}) P_N(\vec{y}\,|\,\vec{x}, s_0)^{\frac{1}{1+\varrho}} \right]^{1+\varrho} \tag{2.37}$$

In order to put this result in a more easily interpretable form, we define the transmission rate of the code to be

$$R = (\ln M)/N . \tag{2.38}$$

If B channel symbols per second are transmitted, then $RB/\ln 2$ is the number of binary digits per second entering the encoder and is the usual rate referred to by communication engineers. It should be emphasized that this rate has nothing to do with the mutual information being transmitted. We can then rewrite (2. 37) in the following form.

$$P_{e,m}(s_0) \leq 4\, A^{1+\varrho} \exp\left\{ -N \left[\max_{0 \leq \varrho \leq 1} - \varrho R + E_{0,N}(\varrho) \right] \right\} \tag{2.39}$$

$$E_{0,N}(\varrho) = \max_{\vec{Q}_N} \min_{s_0} E_{0,N}(\varrho, \vec{Q}_N, s_0) \tag{2.40}$$

$$E_{0,N}(\varrho, \vec{Q}_N, s_0) = -\frac{1}{N} \ln \sum_{\vec{y}} \left[\sum_{\vec{x}} Q_N(\vec{x}) P_N(\vec{y}\,|\,\vec{x}, s_0)^{\frac{1}{1+\varrho}} \right]^{1+\varrho} \tag{2.41}$$

This expression is somewhat messy to analyze, but if we consider the $Q_N(\vec{x})$ that yields $\underline{C}_N$ and the s_0 that yields $\underline{C}_N$, we can make some statements easily. First, the function $E_{0,N}(\varrho, \vec{Q}_N, s_0)$ is convex $\cap$ in ϱ and therefore we get the following parametric equations in ϱ.

$$\max_{0 \le \varrho \le 1} \left[-\varrho R + E_{0,N}(\varrho, \vec{Q}_N, s_0)\right] = -\varrho \frac{\partial E_{0,N}(\varrho, Q_N, s_0)}{\partial \varrho} + E_{0,W}(\varrho, \vec{Q}_N, s_0) \tag{2.42}$$

$$R = \frac{\partial E_{0,N}(\varrho, \vec{Q}_N, s_0)}{\partial \varrho}. \tag{2.43}$$

By evaluating this partial derivative at $\varrho = 0$, we find that at this point $R = \underline{C}_N$ and that the exponent given by the maximum in (2.42) is zero. As ϱ increases, R decreases and the exponent increases. Finally, for $R < R_{crit}$ when R_{crit} is defined as $\partial E_{0,N}(\varrho, \vec{Q}_N, s_0) / \partial \varrho \,|_{\varrho = 1}$, the above equations are no longer valid and we have

$$\max_{0 \le \varrho \le 1} \left[-\varrho R + E_{0,N}(\varrho, \vec{Q}_N, s_0)\right] = E_{0,1}(1, \vec{Q}_N, s_0) - R \tag{2.44}$$

These equations are sketched in figure 2.2.

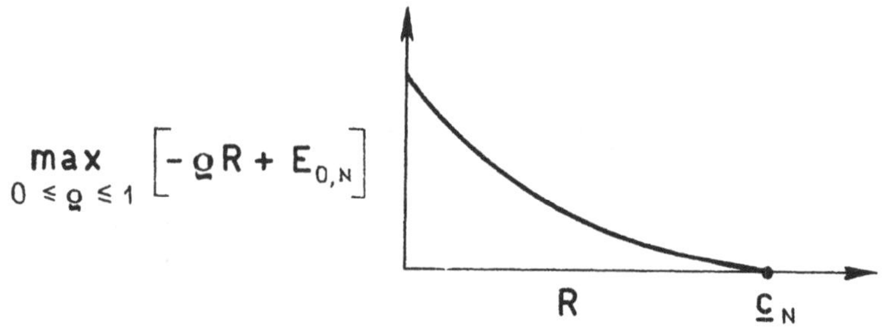

Figure 2.2

Since the exponent shown in Figure 2. 2. is a lower bound to the exponent in brackets in (2. 39), and since $\underline{C}_N$ approaches $\underline{C}$ with increasing N, it is reasonable to assume that the error probability must decrease exponentially with N for large N. Actually, this result is not obvious, since we are not yet sure that the exponent at values of $R < \underline{C}$ does not decrease with increasing N. The next theorem, which is due to Yudkin and proved in section 5. 9 of Gallager, shows that the bracketed term in (2. 39) does indeed stay bounded above 0 for sufficiently large N and for all rates less than $\underline{C}$.

Theorem : The sequence $E_{0,N}(\varrho)$ approaches a limit in N,

$$E_{0,\infty}(\varrho) = \lim_{N} E_{0,N}(\varrho) = \sup_{N} \left[E_{0,N}(\varrho) - \frac{\varrho \ln A}{N} \right] \tag{2. 45}$$

Moreover, the random coding exponent, defined by the function

$$E_r(R) = \max_{0 \leq \varrho \leq 1} E_{0,\infty}(\varrho) - \varrho R \tag{2. 46}$$

is strictly positive for $0 \leq R < \underline{C}$, strictly decreasing in R over this region, and convex $\cup$. Finally, for any $\epsilon > 0$, there exists an $N(\epsilon)$ such that for all $N \geq N(\epsilon)$, and all numbers of code words M, codes exist for which

$$P_{e,m}(s_0) \leq \exp\left[-N(E_r(R) - \epsilon)\right] \tag{2. 47}$$

for each code word m and each initial state s_0.

The above theorem is very general and yields a very powerful result. However, the problem of calculating the random coding exponent is rather formidable. In the special case of a discrete memoryless channel, this result specializes, and $E_r(R)$ can be calculated directly in terms of the transition probabilities without going through any limiting operation in N. Also, for discrete memoryless channel, lower bounds to error probability exist (see Shannon, Gallager, and Berlekamp, Information and Control , pages 65 to 103, 1967) which show that at rates greater than R_{crit} no codes exist for which the error probability decreases with N faster than exponentially with the exponent $E_r(R)$. For rates less than R_{crit}, tighter bounds on error probability can be derived (see Gallager, section 5. 7).

The major reason why these results on error probability for finite state channels are difficult to work with arises because in the general case there is a state process going on which is not observable to the input and output and thus the same input output sequence pairs can arise from many different state sequences. This is a very familiar problem in information theory and is seen most clearly in the case of a source sequence derived from a Markov chain. In this case, the source letters are emitted with given probabilities depending on the state of the Markov chain, but the states can not be determined from knowledge of the previous state and the source output. For this type of situation, no way is known to even find the entropy of the source without looking at a limit of longer and longer sequences.

There are two known situations where we can get tractable results on these exponents. The first is the case where the receiver can track the state of the channel, or mathematically, where s_n is a deterministic function of y_n and s_{n-1}. In the course, this was done using the weaker condition that the input and output together specify the present state given the previous state. Unfortunately, the author has since found an error in that derivation and it now appears that this result is simply wrong. Assuming now that the receiver can track the state, we have

$$P_N(\vec{y},\vec{s} \mid \vec{x}, s_0) = \begin{cases} P_N(\vec{y} \mid \vec{x}, s_0)\,; & \text{for } \vec{s} \text{ determined by } \vec{y}, s_0 \\ 0\,; & \text{otherwise} \end{cases} \qquad (2.48)$$

Using this relation, (2. 41) can be rewritten as

$$E_{0,N}(\rho, \vec{Q}_N, s_0) = -\frac{1}{N} \ln \sum_{\vec{y},\vec{s}} \left[\sum_{\vec{x}} Q_N(\vec{x}) P_N(\vec{y},\vec{s} \mid \vec{x}, s_0)^{\frac{1}{1+\rho}} \right]^{1+\rho} . \qquad (2.49)$$

To verify this, we observe that for each $\vec{y}$, there is only one $\vec{s}$ in the summation for which the term in brackets is non-zero, and for that term we can use (2. 48). This in fact was where the error occured in the derivation assuming that the state was determined by the input and output together. Now we assume that in the ensemble of codes the input letters are independently chosen, so that

(2. 50) $$Q_N(\vec{x}) = \prod_{n=1}^{N} Q(x_n)$$

Using this relation and (2. 2), we rewrite (2. 49) as

(2. 51) $$E_{0,N}(\varrho,\vec{Q}_N,s_0) = -\frac{1}{N}\ln \sum_{\vec{y},\vec{s}} \left[\sum_{\vec{x}} \prod_{n=1}^{N} Q(x_N)P(y_n,s_n|x_n,s_{n-1})^{\frac{1}{1+\varrho}}\right]^{1+\varrho}.$$

Summing over each input letter separately we obtain

(2. 52) $$E_{0,N}(\varrho,Q_N,s_0) = -\frac{1}{N}\ln \sum_{\vec{y},\vec{s}} \left\{\prod_{n=1}^{N} \sum_{x_n} Q(x_n)P(y_n,s_n|x_n,s_{n-1})^{\frac{1}{1+\varrho}}\right\}^{1+p}.$$

We can then take the product sign outside the braces and sum over each output letter separately, obtaining

(2. 53) $$E_{0,N}(\varrho,\vec{Q}_N,s_0) = -\frac{1}{N}\ln \sum_{\vec{s}} \prod_{n=1}^{N} \sum_{y_n} \left[\sum_{x_n} Q(x_n)P(y_n,s_n|x_n,s_{n-1})^{\frac{1}{1+\varrho}}\right]^{1+\varrho}.$$

Now define

(2. 54) $$\alpha(s_{n-1},s_n) = \sum_{y_n}\left[\sum_{x_n} Q(x_n)P(y_n,s_n|x_n,s_{n-1})^{\frac{1}{1+\varrho}}\right]^{1+\varrho}$$

(2. 55) $$E_{0,N}(\varrho,\vec{Q}_N,s_0) = -\frac{1}{N}\ln \sum_{\vec{s}} \prod_{n=1}^{N} \alpha(s_{n-1},s_n).$$

Finally, define the A by A matrix

(2. 56) $$[\alpha] = \begin{bmatrix} \alpha(0,0) & \alpha(0,1) & \dots & \alpha(0,A-1) \\ \vdots & & & \\ \alpha(A-1,0) & & \dots & \alpha(A-1,A-1) \end{bmatrix}$$

Let $e(s_0)$ be an A dimensional row vector with a 1 in the s_0^{th} position and zeros elsewhere and let $1]$ be an A dimensional column vector of all 1's. Then,

$$E_{0,N}(\varrho, \vec{Q}_N, s_0) = -\frac{1}{N} \ln\left\{e(s_0)[\alpha]^N 1]\right\}. \qquad (2.57)$$

This equation can be verified by checking it for $N = 2$ and then using induction. This matrix has all nonnegative terms, and assuming that each state can be reached from each other state in a finite number of transitions, we can apply the Frobenius theorem which states that $[\alpha]$ has a positive real eigenvalue, larger or equal in magnitude to all the other eigenvalues, and that this eigenvector has a right eigenvector of positive elements. Letting $\lambda(\varrho)$ be this eigenvalue and $\upsilon]$ be the eigenvector, we have

$$[\alpha]^N \upsilon] = [\alpha]^{N-1} \lambda(\varrho)\upsilon] = \ldots = [\lambda(\varrho)]^N \upsilon]. \qquad (2.58)$$

Now let υ_{max} and υ_{min} be the largest and smallest term in the vector $\upsilon]$. We see that $1] \leq \upsilon] / \upsilon_{min}$, and thus

$$[\alpha]^N 1] \leq [\alpha]^N \upsilon] / \upsilon_{min} \leq \lambda^N(\varrho)\upsilon] / \upsilon_{min}. \qquad (2.59)$$

Finally the premultiplication by $e(s_0)$ will pick one term out of this vector, so that

$$e(s_0)[\alpha]^N 1] \leq \lambda^N(\varrho)\upsilon_{max} / \upsilon_{min}. \qquad (2.60)$$

Going to the limit as $N \to \infty$, the v_{max} and v_{min} terms disappear. Also we can choose $Q(x)$ to maximize the eigenvalue $\lambda(\varrho)$. We then have

$$E_{0,\infty}(\varrho) \geq -\ln \lambda(\varrho) \tag{2.61}$$

This inequality will become an equality if a product distribution for Q_N actually maximizes $E_{0,N}$. Certainly this will be the case if the same $Q(x)$ maximizes $\alpha(s_{n-1}, s_n)$ for all values of s_{n-1} and s_n.

The second situation where we can get tractable results on these exponents is that where for an independent, equally distribution on the inputs, the quantity

$$\sum_{n} Q(x_n) P(y_n, s_n | x_n, s_{n-1})^{\frac{1}{1+\varrho}} \tag{2.62}$$

is independent of the value of y_n. Such a model is sometimes appropriate where the inputs and outputs are equally spaced phases on a carrier and the memory is in the form of a slowly varying phase difference between input and output. Using the inequality (2. 34), we have

$$E_{0,N}(\varrho, \vec{Q}_N, s_0) \geq -\frac{1}{N} \ln \sum_{\vec{y}} \left\{ \sum_{\vec{s}} \sum_{\vec{x}} \prod_{n=1}^{N} Q(x_n) P(y_n, s_n | x_n, s_{n-1})^{\frac{1}{1+\varrho}} \right\}^{1+\varrho} . \tag{2.63}$$

We can sum over x as before and then define

$$\alpha(s_{n-1}, s_n) = \sum_{x_n} Q(x_n) P(y_n, s_n | x_n, s_{n-1})^{\frac{1}{1+\varrho}} \tag{2.64}$$

Recalling that this quantity is independent of y_n, we obtain

$$E_{0,N}(\varrho, \vec{Q}_N, s_0) \geq -\frac{1}{N} \ln \sum_{\vec{y}} \left\{ \sum_{\vec{s}} \alpha(s_{n-1}, s_n) \right\}^{1+\varrho}. \qquad (2.65)$$

We can define the matrix $[\alpha]$ as before and its largest eigenvalue $\lambda(\varrho)$ and eigenvector $v]$. As before, we obtain

$$E_{0,N}(\varrho, \vec{Q}_N, s_0) \geq -\frac{1}{N} \ln \left\{ \sum_{\vec{y}} \left[\lambda^N(\varrho)\right]^{1+\varrho} \left[\frac{v_{max}}{v_{min}}\right]^{1+\varrho} \right\}. \qquad (2.66)$$

Since there are J^N different output sequences y, the sum over y just introduces a multiplicative factor of J^N. Our final result for this case is then

$$E_{0,\infty}(\varrho) \geq -(1+\varrho) \ln \lambda(\varrho) - \ln J. \qquad (2.67)$$

Source Coding with a Distortion Measure

Many of the sources encountered in communication theory produce waveforms, pictures, or a sequence of analogue variables as their output. These can not be recreated exactly at the destination and often much of the detail in the source output is irrelevant. Thus, in defining how effective the communication is, it is necessary to define a distortion measure between the source output and the received message and to evaluate the effectiveness of the communication in terms of the expected value of that distortion.

For simplicity of notation and terminology, we restrict ourselves here to discrete time sources. Thus, the source output is a stationary sequence of symbols, say ..., $u_{-1}, u_0, u_1, \ldots$ where for each time unit i, $u_i \in U$ where U is the alphabet of source letters. The alphabet might be finite, or countably infinite, or the real line, or a suitably restricted set of functions of one or more variables. There is also an alphabet V of destination letters which might or might not be the same as the source alphabet. The source is defined by a probability measure on source sequences and by a distortion measure $D(u,v)$ $u \in U, v \in V$, which is a real measurable function on $U \times V$. We assume here a suitable specification of measurable subsets of U, V, and the extension to $U \times V$ and to sequences of symbols. If $\vec{u} = (u_1, \ldots, u_n)$ is a sequence of source symbols and $\vec{v} = (v_1, \ldots, v_n)$ is the corresponding sequence of destination symbols, we define the distortion per symbol between $\vec{u}$ and $\vec{v}$ as

$$D_n(\vec{u}, \vec{v}) = \frac{1}{n} \sum_{i=1}^{n} D(u_i, v_i). \tag{3.1}$$

One might argue that it is foolish, from the standpoint of real sources, to restrict D_n as in (3.1) without also restricting the source symbols to be statistically independent. This is true, but it simplifies the mathematics considerably. A more general assumption than (3.1) is considered in section 9.8 of Gallager, but most of the literature uses (3.1).

We shall now define a function $R(D^*)$ called the rate distortion function for the source. It depends only on the source probability measure and distortion function and is a decreasing, convex U function of D^*. Its significance comes from the coding theorem for sources with a distortion measure. Stated imprecisely this theorem is as follows : suppose we wish to transmit the source output over a channel and achieve an average distortion of D^* or less. Then, this is possible, by sufficiently complex processing at the source and destination, if the capacity of the channel, in nats/source symbol, is greater than $R(D^*)$, and is impossible, no matter what processing is used, if the capacity is less than $R(D^*)$. Even stronger, if the capacity

C is greater than $R(D^*)$ than average distortion D^* can be achieved by first converting the source output into a binary stream, using $R'/\ln 2$ binary digits per source symbol, where $R(D^*) < R' < C$.

In order to define $R(D^*)$, we first define the rate distortion function $R_n(D^*)$ for sequences of length n.

(3. 2) $$R_n(D^*) = \inf_{P_{\vec{u},\vec{v},n} \in \mathcal{P}_n(D^*)} \frac{1}{n} I(U^n; V^n).$$

Here $P_{\vec{u},\vec{v},n}$ is a joint probability measure over a sequence of n inputs and n outputs, and $I(U^n, V^n)$ is the average mutual information (in nats) for that joint probability measure. The class $\mathcal{P}_n(D^*)$ over which the infimum is taken is the set of joint proabability measures whose marginal distribution on the input sequence is the Q_N specified by the source probability measure and for which $E\left[D(\vec{u},\vec{v})\right] \leq D^*$. We then define

(3. 3) $$R(D^*) = \liminf_{n} R_n(D^*).$$

It can be shown (see section 9. 8 of Gallager) that $R(D^*)$ is also $\lim R_n(D^*)$ and is also $\inf R_n(D^*)$.

In order to gain insight into the behavior of $R(D^*)$, we first make the assumption that $D(u,v)$ satisfies

$$\inf_{v \in V} D(u,v) = 0 \,; \quad \text{all } u \in U.$$

This really entails no loss of generality, since for an arbitrary distortion measure, $D'(u,v)$, we can define

$$D(u,v) = D'(u,v) - \inf_{v} D'(u,v) \tag{3.4}$$

Since $D(u,v) - D'(u,v)$ is a function only of u, the expected value of this difference is independent of the processing done between source and destination, and thus irrelevant to our purposes. We also say that $R_n(D^*)$ is undefined if the set $\mathcal{P}_n(D^*)$ is empty, so that $R_n(D^*)$ is undefined for $D^* < 0$ and might or might not be undefined for $D^* = 0$ depending on the source. It is clear that if $R_n(D^*)$ is defined for one value of D^*, it is defined for all larger D^* since the set $\mathcal{P}_n(D^*)$ increases with D^*.

Theorem 1 : For each $n > 0$, $R_n(D^*)$ is nonnegative, non-increasing in D^*, and convex $\cup$ over the region of D^* where it is defined.

Proof : The nonnegativity follows from the nonnegativity of mutual information and the non-increasing property follows from the fact that the constraint set $\mathcal{P}_n(D^*)$ is non-decreasing with D^*. For the convexity, we can set $n = 1$ without loss of generality (for larger n, just consider the ensemble of sequences of length n as one ensemble U). For arbitrary $\epsilon > 0$ and arbitrary D_1^* and D_2^*, let $P'_{u,v}$ and $P''_{u,v}$ be in $\mathcal{P}_1(D_1^*)$ and $\mathcal{P}_1(D_2^*)$ respectively with mutual information $f(P'_{u,v})$ and $f(P''_{u,v})$ at most $R_1(D_1^*)+\epsilon$ and $R_1(D_2^*)+\epsilon$ respectively. For any given λ, $0<\lambda<1$, let

$$P_{u,v} = \lambda P'_{u,v} + (1-\lambda) P''_{u,v} \tag{3.5}$$

Then $P_{u,v} \in \mathcal{P}_1(\lambda D_1^* + (1-\lambda) D_2^*)$. Finally, if $P_{v|u}$ is the transition probability measure for $P_{u,v}$, then

$$P_{v|u} = \lambda P'_{v|u} + (1-\lambda) P''_{v|u} \tag{3.6}$$

Since mutual information is convex $\cup$ in the transition probabilities,

$$f(P_{u,v}) \le \lambda f(P'_{u,v}) + (1-\lambda) f(P''_{u,v}) \le$$

$$\le \lambda R_1(D_1^*) + (1-\lambda) R_1(D_2^*) + \epsilon .$$

Thus

$$R_1(\lambda D_1^* + (1-\lambda) D_2^*) \le \lambda R_1(D_1^*) + (1-\lambda) R_1(D_2^*) + \epsilon . \tag{3.7}$$

Since $\epsilon > 0$ was arbitrary, this establishes the convexity.

It follows as an easy corollary that $R(D^*)$ is also nonnegative, non-increasing and convex $\cup$. To establish the convexity, however, one must use the fact that $R(D^*) = \lim R_n(D^*)$.

In most of what follows, we restrict our attention to memoryless sources; i. e. sources for which successive letters are statistically independent and identically distributed.

Theorem : For a memoryless source

$$R(D^*) = R_1(D^*) \tag{3.8}$$

Proof : For arbitrary $\epsilon > 0$, let $P_{u,v}$ be a probability measure on U, V for which $E[D(u,v)] \leq D^*$ and for which $I(U;V) \leq R_1(D^*) + \epsilon$. The nth order product measure formed from $P_{u,v}$ yields at most average distortion D^*, and average mutual information per letter equal to $I(U;V)$. Thus

$$R_n(D^*) \leq R_1(D^*) + \epsilon \quad \text{for every } \epsilon > 0. \tag{3.9}$$

Next, for arbitrary n, let $P_{\vec{u},\vec{v},n}$ be in $\mathcal{P}_n(D^*)$ and satisfy

$$\frac{1}{n} I(U^n; V^n) \leq R_n(D^*) + \epsilon. \tag{3.10}$$

Denoting U^n by the joint ensemble $U_1, \ldots, U_n$ and V^n by $V_1, \ldots, V_n$, we obtain

$$I(U^n; V^n) = \sum_{\ell=1}^{n} I(U_\ell; V^n | U_{\ell-1}, \ldots, U_1) \tag{3.11}$$

$$I(U_\ell; V^n | U_{\ell-1}, \ldots, U_1) = I(U_\ell; V^n, U_{\ell-1}, \ldots, U_1) -$$

$$- I(U_\ell; U_{\ell-1}, \ldots, U_1) = I(U_\ell; V^n, U_{\ell-1}, \ldots, U_1)$$

$$\geq I(U_\ell; V_\ell) \tag{3.12}$$

(3.13)
$$I(U^n; V^n) \geq \sum_{\ell=1}^{n} I(U_\ell ; V_\ell)$$

Let $\bar{D}_\ell$ be the average distortion on the ℓth letter. Since $P_{\vec{u},\vec{v},n} \in \mathcal{P}_n(D^*)$, we can use (3.1) to see that

(3.14)
$$\frac{1}{n} \sum_{\ell=1}^{n} \bar{D}_\ell \leq D^*.$$

Also from the definition of $R_1(D^*)$, we see that

(3.15)
$$I(u_\ell ; V_\ell) \geq R_1(\bar{D}_\ell).$$

Combining (3.10), (3.13) and (3.15)

(3.16)
$$R_n(D^*) + \epsilon \geq \frac{1}{n} \sum_{\ell=1}^{n} R_1(\bar{D}_\ell).$$

Using the convexity and non-increasing property of R_1,

(3.17)
$$R_n(D^*) + \epsilon \geq R_1 \left(\frac{1}{n} \sum_{\ell=1}^{n} \bar{D}_\ell\right) \geq R_1(D^*).$$

Combining (3.8) and (3.17), $R_n(D^*) = R_1(D^*)$ for all n, completing the proof.

Before going on to prove the source coding theorem, we give some results on how to calculate the rate distortion function for memoryless sources. Since the source is memoryless, $R(D^*) = R_1(D^*)$, and the definition of $R_1(D^*)$ in (3.1) suggests that it can be found by a Lagrange multiplier technique. Thus we define $R_0(\varrho, P)$ as the sum of the mutual information

mation with the joint measure P and ϱ times the average distortion with the joint measure P,

$$R_0(\varrho, P) = \int \ln\left[\frac{dP_{uv}}{dP_{u\times v}}\right] dP_{u,v} + \varrho \int D(u,v)\, dP_{u,v}$$

$$= \int \ln\left[\frac{dP_{uv}}{dP_{u\times v}}(u,v)\, e^{\varrho D(u,v)}\right] dP_{u,v} \qquad (3\ 18)$$

In (3 18), $dP_{uv} / dP_{u\times v}$ is the Radon-Nikodym derivative of the joint measure on the UV space with respect to the corresponding product measure. As usual, we take this integral to be infinite if P_{uv} is not absolutely continuous with respect to $P_{u\times v}$. We then define

$$R_0(\varrho) = \inf R_0(\varrho, P) \qquad (3.19)$$

where the infimum is over all measures $P_{u,v}$ on the joint space with the given marginal probability measure on the source space U.

For discrete spaces, $P_{u,v}$ is uniquely determined by the source probability assignment, $Q(u)$ and by a transition probability assignment $P(v|u)$ for all $u \in U$, $v \in V$, and (3. 18) becomes

$$R_0(\varrho, P) = \sum_{u,v} Q(u) P(v|u) \ln \frac{P(v|u)}{\sum_{u'} Q(u') P(v|u')} +$$

$$+ \varrho \sum_{u,v} Q(u)\, P(v|u)\, D(u,v). \qquad (3.20)$$

The connection between $R_0(\varrho)$ and $R(D^*)$ is shown in Fig. 3.1. Each $P_{u,v}$ gives rise to an expected distortion and an average mutual information. Plotting distortion on the horizontal axis and information on the vertical axis, each $P_{u,v}$ corresponds to a point in the plane. The curve $R(D^*)$, by definition, is the infimum of this set of points. On the other hand, for each $\varrho \geq 0$, the straight line $R_0(\varrho) - \varrho D^*$ has slope $-\varrho$, intersects the vertical axis at $R_0(\varrho)$, and each point in the plane corresponding to some $P_{u,v}$ lies on or above each of these lines, but approach each of those lines arbitrarily closely. Using the convexity of $R(D^*)$, it is seen from this that for each ϱ, the line $R_0(\varrho) - \varrho D^*$ is tangent to the curve $R(D^*)$. Since a convex curve is determined by its tangents, finding $R_0(\varrho)$ is equivalent to finding $R(D^*)$. We can usually express $R(D^*)$ parametrically in terms of ϱ by

$$D^* = \frac{dR_0(\varrho)}{d\varrho}$$

(3.21) $$R(D^*) = R_0(\varrho) - \frac{dR_0(\varrho)}{d\varrho}$$

We see from the above that it is not particularly difficult, especially using numerical means, to find $R(D^*)$ from $R_0(\varrho)$. The difficult problem is finding $R_0(\varrho)$, and as can be seen from (3.20), this is not a trivial task even for discrete finite memoryless sources.

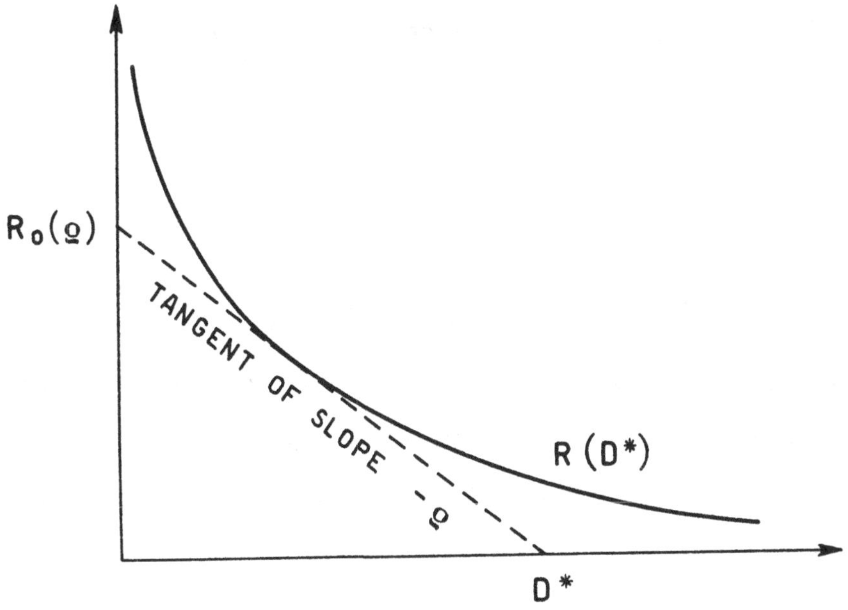

Geometric Relation Between $R(D^*)$ and $R_0(\varrho)$
Figure 3. 1

In the results to follow, we shall not only simplify the mechanism of calculating $R_0(\varrho)$ but also establish general upper and lower bounds to $R_0(\varrho)$ (and hence to $R(D^*)$) and also develop results that are useful in theoretical work.

First let (ω) be an arbitrary probability measure on the destination space V and define

$$R_0(\varrho, P, \omega) = \int dP_{uv} \, \ln\left[\frac{dP_{uv}}{d(P_u x \omega)} e^{\varrho D(u,v)}\right] \qquad (3.22)$$

$$F(\omega) = -\int dP_u \, \ln \int_v d\omega \; e^{-\varrho D(u,v)}. \qquad (3.23)$$

As before, we take $R_0(\varrho, P, \omega)$ to be infinite if P_{uv} is not absolutely continuous with respect to the product measure $P_u \times \omega_v$.

Theorem

$$R_0(\varrho) = \inf_{P,\omega} R_0(\varrho, P, \omega) = \inf_{\omega} F(\omega) \tag{3.24}$$

when the infimum over P, as before, is over all joint probability measures on U, V for which the marginal on U is the given source measure and the infimum on ω is over all probability measures on V.

Proof :

First we establish two subsidiary results

1) $$R_0(\varrho, P) \le R_0(\varrho, P, \omega) \tag{3.25}$$

with equality if ω is the marginal measure on V induced by $P_{u,v}$

2) $$F(\omega) \le R_0(\varrho, P, \omega) \tag{3.26}$$

with equality if P satisfies

$$\frac{dP_{uv}}{d(P_u \times \omega)}(u, v) = \frac{e^{-\varrho D(u,v)}}{\int_{v'} e^{-\varrho D(u,v')}\, d\omega} \quad \text{for all } u \in U \tag{3.27}$$

To establish (3.25), we use the definition of $R_0(\varrho, P)$ and $R_0(\varrho, P, \omega)$ in (3.18) and (3.22) to obtain

$$R_0(\varrho,P) - R_0(\varrho,P,\omega) = \int \left\{ \ln\left[\frac{dP_{u,v}}{dP_{u\times v}}\right] - \ln\left[\frac{dP_{u,v}}{d(P_u \times \omega)}\right]\right\} dP_{u,v}$$

$$= \int \ln\left[\frac{dP_{u,v}}{dP_{u\times v}} \; \frac{d(P_u \times \omega)}{dP_{u,v}}\right] dP_{u,v}. \qquad (3.28)$$

Here we have used the fact that $d(P_u \times \omega)/dP_{u,v}$ is defined and equal to the reciprocal of $dP_{u,v}/d(P_u \times \omega)$ except on a set of $P_{u,v}$ measure zero.

$$R_0(\varrho,P) - R_0(\varrho,P,\omega) = \int_{u,v} \ln\left[\frac{d\omega}{dP_v}\right] dP_{u,v}$$

$$= \int_v \ln\left[\frac{d\omega}{dP_v}\right] dP_v \leq \int_v d\omega - \int_v dP_v = 0 \qquad (3.29)$$

where we have and the inequality $\ln x \leq x-1$. If ω and P_v are identical, then the inequality becomes an equality.

Eq. (3.27) is established in the same way :

$$F(\omega) - R_0(\varrho,P,\omega) =$$

$$= \int dP_{uv} \ln\left[\frac{e^{-\varrho D(u,v)}}{\frac{dP_{u,v}}{d(P_u \times \omega)}(u,v) \quad e^{-\varrho D(u,v')} d\omega}\right]$$

(3.30)

$$\leq \int \frac{d(P_u \times \omega)\, e^{-\varrho D(u,v)}}{e^{-\varrho D(u,v)}\, d\omega} - 1$$

$$= \int_u d P_u - 1 = 0$$

Equality is achieved when the term in brackets in (3. 30) is 1 almost everywhere with respect to $P_{u,v}$, this establishing the conditions for equality in (3. 27).

The proof of the theorem is now quite easy. Given any $P_{u,v}$, we can choose ω to be equal to the marginal probability measure of $P_{u,v}$ on V, and for that ω,

$$R_0(\varrho, P) = R_0(\varrho, P, \omega) \geq F(\omega) \tag{3. 31}$$

Thus

$$R_0(\varrho) = \inf_P R_0(\varrho, P) \geq \inf R_0(\varrho, P, \omega) \geq \inf F(\omega). \tag{3. 32}$$

Conversely, given any ω, we can choose $P_{u,v}$ to satisfy (3. 27), and for that choice of $P_{u,v}$

$$F(\omega) = R_0(\varrho, P, \omega) \geq R_0(\varrho, P) \tag{3. 33}$$

and

$$\inf F(\omega) \geq \inf R_0(\varrho, P, \omega) \geq \inf R_0(\varrho, P) = R_0(\varrho). \tag{3. 34}$$

This completes the proof since (3. 32) and (3. 34) imply (3. 24).

The above theorem is due to Haskell (Trans. I. T., Sept. 1969, p. 525).

In many cases, particularly when the U and V alphabets are infinite, it is difficult to find $F(\omega)$, but $F(\omega)$, for any given ω, still serves as an upper bound on $R_0(\varrho)$, and since $F(\omega)$ is convex $\cup$ in ω, this upper bound is easy to work with. We now go on to find a corresponding lower bound to $R_0(\varrho)$. For an arbitrary measurable function $h(u)$ on the input space, define

$$G(h) = \int_u dP_u \, \ln h(u) . \tag{3.35}$$

Let H be the set of functions

$$H = \left\{ h(u) : \int_u dP_u \, h(u) \, e^{-\varrho D(u,v)} \leq 1 \ ; \ \text{all} \ \ v \in V \right\} . \tag{3.36}$$

Theorem : For all $h(u) \in H$

$$G(h) \leq R_0(\varrho) . \tag{3.37}$$

Proof : It suffices to show that $G(h) \leq F(\omega)$ for each $h \in H$ and each destination measure ω. But

$$G(h) - F(\omega) = \int dP_u \, \ln \left[h(u) \int_v d\omega \, e^{-\varrho D(u,v)} \right] \leq$$

$$\leq \int dP_u \, h(u) \int_v d\omega \, e^{-\varrho D(u,v)} - 1 = \tag{3.38}$$

$$= \int_{v} d\omega \int_{u} dP_u \, h(u) \, e^{-\varrho D(u,v)} - 1 \le 0 .$$

The first inequality is the familiar $\ln x \le x - 1$ and the second comes from the constraint (3. 36).

The next question that arises is whether the supremum of $G(h)$ over $h \in H$ is equal to $R_0(\varrho)$. We first consider the case when the source and destination alphabets are finite. Then $\inf F(\omega)$ exists as a minimum and finding the minimum is the familiar problem of minimizing a convex $\cup$ function of a probability vector. Differentiating $F(\omega)$ with respect to each component $\omega(v)$, we find that necessary and sufficient condition on ω to minimize $F(\omega)$ are that for some constant λ,

(3. 39) $$\sum_u P_u(u) \frac{e^{-\varrho D(u,v)}}{\sum_v \omega(v) e^{-\varrho D(u,v)}} \le \lambda \text{ for each } v \in V$$

with equality for v such that $\omega(v) > 0$. Multiplying each side of (3. 39) by $\omega(v)$ and summing over v, we find that $\lambda = 1$. Finally setting

(3. 40) $$h(u) = \frac{1}{\sum \omega(v) e^{-\varrho D(u,v)}}$$

we see that (3. 39) states that $h(u) \in H$. Next, substituting (3. 40) into (3. 35), we see that $G(h) = F(\omega)$. Thus for finite alphabets

$$\max_{h \in H} G(h) = R_0(\varrho) = \min_{\omega} F(\omega). \tag{3.41}$$

Next consider this case when the source alphabet is finite and the destination alphabet is countable.

Lemma : For a finite source alphabet and countable destination alphabet,

$$\max_{h \in H} G(h) = R_0(\varrho). \tag{3.42}$$

Proof : Let the destination alphabet be represented by the natural numbers and for each M, let

$$H_M = \left\{ h(u) \geq 0 : \sum_u P_u(u) h(u) e^{-\varrho D(u,v)} \leq 1 ; 1 \leq v \leq M \right\}.$$

Let $h_M(u)$ be the $h(u)$ that maximizes $G(h)$ over $h \in H_M$. Note that $G(h_M)$ is equal to $R_0(\varrho)$ for a source and distortion measure identical to our original source with the restriction that only the outputs 0 to M are available. Thus we have

$$G(h_M) \geq R_0(\varrho) ; \text{ all } M \tag{3.43}$$

Finally we have $H_1 \supset H_2 \supset \dots \supset H_M \supset \dots$. Also each H_M is closed and bounded(*), and thus the sequence $h_1(u), h_2(u), \dots$

(*) If infinite distortions are allowed, then H_M is only bounded for M large enough so that for each u, there is some $v \leq M$ with $D(u,v) < \infty$. If no such M exists $\max G(h) = \infty = R_0(\varrho)$.

has a cluster point, say $h_\infty(u)$. From (3. 43) we see that $G(h_\infty) \geq R_0(\varrho)$. Also, since h_∞ is a cluster point of $\{h_M\}$, we see that

$$\sum_u P_u(u) h_\infty(u) e^{-\varrho D(u,v)} \leq 1 \quad \text{for all } v.$$

Thus $h_\infty \in H$ and $G(h_\infty) \geq R_0(\varrho)$, which with (3. 37) completes the proof.

Theorem : For a source with countable input and output alphabets,

$$\sup_{h \in H} G(h) = R_0(\varrho) \tag{3.44}$$

Proof : Assume first that $R_0(\varrho)$ is finite. For notational convenience, let $\vec{P}$ in $R_0(\varrho, \vec{P})$ denote a transition probability assignment rather than the joint probability assignment. Choose $\vec{P}$ so that $R_0(\varrho, \vec{P})$ is finite and for arbitrary $\epsilon > 0$ choose N large enough to satisfy the following conditions :

$$H(\alpha) = -\alpha \ln \alpha - (1-\alpha) \cdot \ln(1-\alpha) \leq \epsilon \tag{3.45}$$

where

$$\alpha = \sum_{u=0}^{N} Q_u(u)$$

$$\sum_{u=0}^{N} Q_u(u) P(v|u) \ln \frac{P(v|u) e^{-\varrho D(u,v)}}{\sum_{u'} Q_u(u') P(v|u')} \geq$$

$$\geq R_0(\varrho, P) - \epsilon . \tag{3.46}$$

Now define

$$Q'(u) = \begin{cases} Q_u(u)/\alpha \; ; & 0 \le u \le N \\ 0 \; ; & u > N \end{cases} \tag{3.47}$$

$$Q''(u) = \begin{cases} Q_u(u)/(1-\alpha) \; ; & u > N \\ 0 \; ; & u \le N \end{cases} \tag{3.48}$$

Then $\vec{Q}'$ and $\vec{Q}''$ are each probability assignments over U and the source distribution $\vec{Q}$ satisfies $\vec{Q} = \alpha\vec{Q}' + (1-\alpha)\vec{Q}''$. Recalling our proof of the convexity of mutual information with respect to the input probabilities, we have

$$\alpha F(\vec{Q}',\vec{P}) + (1-\alpha)F(\vec{Q}'',\vec{P}) \le F(\vec{Q},\vec{P}) \le$$

$$\le \alpha F(\vec{Q}',\vec{P}) + (1-\alpha)F(\vec{Q}'',\vec{P}) + H(\alpha). \tag{3.49}$$

Let $R_0'(\varrho,\vec{P})$ and $R_0''(\varrho,\vec{P})$ be the function $R_0(\varrho,\vec{P})$ for the source distribution $\vec{Q}'$ and $\vec{Q}''$ respectively. From the definition of $R_0(\varrho,\vec{P})$, (3.49) is equivalent to

$$\alpha R_0'(\varrho, \vec{P}) + (1-\alpha) R_0''(\varrho, \vec{P}) \le$$

(3.50) $$\le R_0(\varrho, \vec{P}) \le \alpha R_0'(\varrho, \vec{P}) + (1-\alpha) R_0''(\varrho, \vec{P}) + H(\alpha).$$

Next we want to show that $R_0''(\varrho, \vec{P})$ is small, and to use this to allow us to work with only the first $N+1$ inputs. Using (3.47), we have

$$\alpha R_0'(\varrho, P) = \alpha \sum_{u=0}^{N} \sum_{v} \frac{Q_u(u)}{\alpha} P(v|u) \ln \frac{P(v|u) e^{-\varrho D(u,v)}}{\sum_{u'=0}^{N} \frac{Q_u(u')}{\alpha} P(v|u')}$$

Removing α from the $\ln$ term, then lower bounding by extending the sum on u' to infinity, and then using (3.46), we obtain

$$\alpha R_0'(\varrho, \vec{P}) \ge \alpha \ln \alpha + R_0(\varrho, \vec{P}) - \epsilon$$

(3.51) $$\ge R_0(\varrho, \vec{P}) - 2\epsilon$$

where we have used (3.45). Combining the left side of (3.50) with (3.51),

(3.52) $$(1-\alpha) R_0''(\varrho, \vec{P}) \le 2\epsilon .$$

Not let $\vec{P}'$ be any new transition assignment in

$$P'(v|u) = \begin{cases} P(v|u) ; & u > N \\ \text{arbitrary} ; & u \le N \end{cases}$$

Since $\vec{P}'$ and $\vec{P}$ are the same as far as the source $\vec{Q}''$ is concerned, $R_0''(\varrho, P') = R_0''(\varrho, \vec{P})$. Using the right hand side of

(3. 50), with $\vec{P}'$ in place of $\vec{P}$, we have

$$\alpha R_0'(\varrho, P') \geq R_0(\varrho, \vec{P}') - (1-\alpha) R_0''(\varrho, \vec{P}) - H(\alpha)$$

$$\geq R_0(\varrho) - 3\epsilon \tag{3. 53}$$

where we have used (3. 52), (3. 45), and the definition of $R_0(\varrho)$ Since the source $\vec{Q}'$ has a finite input alphabet, we can choose $h'(u)$, $0 \leq u \leq N$ such that

$$\sum_{u=0}^{N} Q'(u) \ln h'(u) = \inf_{\vec{P}'} R_0'(\varrho, \vec{P}') \tag{3. 54}$$

$$\sum_{u=0}^{N} Q'(u) h'(u) e^{-\varrho D(u,v)} \leq 1 \text{ ; all } v. \tag{3. 55}$$

Finally, choose $h(u)$ as

$$h(u) = \begin{cases} h'(u) \text{ ; } & 0 \leq u \leq N \\ 1 \text{ ; } & u > N \end{cases}$$

Then,

$$\sum_{u} Q_u(u) h(u) e^{-\varrho D(u,v)} = \alpha \sum_{u=0}^{N} Q'(u) h'(u) e^{-\varrho D(u,v)}$$

$$+ (1-\alpha) \sum_{u=N+1}^{\infty} Q''(u) e^{-\varrho D(u,v)}$$

$$\leq \alpha + (1-\alpha) = 1 \text{ ; all } v. \tag{3. 56}$$

Thus, $h \in H$. Also

$$\sum_{u} Q_u(u) \ln h(u) = \alpha \sum_{u=0}^{N} Q'(u) \ln h'(u) = \alpha \inf_{\vec{P}} R_0'(\varrho, \vec{P})$$

$$\geq R_0(\varrho) - 3\epsilon \tag{3. 57}$$

Since $\epsilon > 0$ is arbitrary and $G(h) \leq R_0(\varrho)$, this completes the proof.

We now show how to apply these results in two important applications. First consider a discrete finite source U of M letters with a distortion measure that counts all errors as equally serious, i.e.

$$D(u,v) = \begin{cases} 0 \; ; & u = v \\ 1 \; ; & u \neq v \end{cases} \tag{3.58}$$

The set H for this source is then

$$\begin{aligned} H &= \Big\{ h(u) : Q_u(v)h(v) + \sum_{u \neq v} Q_u(u)h(u)e^{-\varrho} \leq 1 \; ; \text{ all } v \Big\} \\ &= \Big\{ h(u) : Q_u(v)h(v)\big[1 - e^{-\varrho}\big] + \sum_{u} Q_u(u)h(u)e^{-\varrho} \leq 1 ; \text{ all } v \Big\}. \end{aligned}$$

These constraints are all satisfied by choosing $Q_u(u)\,h(u)$ independent of u i.e.

$$h(u) = \frac{1}{Q_u(u)\big[1 + (M-1)e^{-\varrho}\big]} \tag{3.59}$$

$$R_0(\varrho) \geq H(U) - \ln\big[1 + (M-1)e^{-\varrho}\big] \tag{3.60}$$

$$\begin{aligned} R(D^*) &= \max_{\varrho} \big[R(\varrho) - \varrho D^*\big] \\ &\geq H(U) - \ln\big[1 + (M-1)e^{-\varrho}\big] - \varrho D^* \end{aligned} \tag{3.61}$$

for all $\rho \geq 0$. The tightest bound is achieved for

$$\rho = \ln\left[(M-1)(1-D^*)/D^*\right] \tag{3.62}$$

$$R(D^*) \geq H(U) - H(D^*) - D^* \ln(M-1) \tag{3.63}$$

where $H(D^*) = -D^* \ln D^* - (1-D^*) \ln(1-D^*)$.

Finally we have seen from (3.40) that (3.60) will be satisfied with equality if a probability assignment $\omega(v)$ exists for which

$$h^{-1}(u) = \sum_v \omega(v) e^{-\rho D(u,v)} \; ; \text{ all } u \tag{3.64}$$

$$h^{-1}(u) = \omega(u)\left[1-e^{-\rho}\right] + \sum_v \omega(v) e^{-\rho} ; \text{ all } u . \tag{3.65}$$

Combining (3.65) and (3.59)

$$\omega(u) = \frac{Q_u(u)\left[1+(M-1)e^{-\rho}\right] - e^{-\rho}}{1-e^{-\rho}} \tag{3.66}$$

We see that $\omega(u)$ will be nonnegative for each u if

$$\min_u Q_u(u) \geq \frac{e^{-\rho}}{1+(M-1)e^{-\rho}} \tag{3.67}$$

Combining this with (3.62), we find that (3.63) is satisfied with equality for D^* satisfying $0 \leq D^* \leq (M-1) \min Q_u(u)$.

For this particular distortion measure, one can go on to find $R(D^*)$ for larger D^* also. That is done in section 9.5 of Gallager but will not concern us here. What does concern us here is the interpretation of (3.63). Suppose the source output is being transmitted over a channel of capacity C nats per source symbol. Suppose that $C \leq H(U)$ and that we choose to satisfy

$$R(D^*) = C \tag{3.68}$$

Since $R(0) = H(U)$ from (3.63), such a D^* exists. The source coding theorem, which we shall soon prove, states that average distortions arbitrarily close to D^* can be achieved through proper coding, but that average distortions less than D^* cannot be achieved. Now consider a sequence of L letters from the source, and let $P_{e,n}$ be the probability that the nth letter is received in error. Using the distortion measure we have defined

$$P_{e,n} = E\left[D(u_n, v_n)\right]$$

$$< P_e > \triangleq \frac{1}{L} \sum_{n=1}^{L} P_{e,n} = E\left[D_L(\vec{u}, \vec{v})\right] \tag{3.69}$$

Thus our result says that the time average error probability, $< P_e >$, mut be at least D^* where $R(D^*) = C$, but need be arbitrarily little more than D^*. Using (3.63), we can restate this to say that $< P_e >$ must satisfy

$$H(< P_e >) + < P_e > \ln(M-1) \geq C - H(U) \tag{3.70}$$

Furthermore, if

$$C \geq H(U) - H\left((M-1) \min Q_u(u)\right) - \min Q_u(u)(M-1) \ln(M-1),$$

then systems exist for which $< P_e >$ is arbitrarily close to the value for which (3.70) is satisfied with equality. Eq. (3.70) is called the converse to the noisy channel conding theorem, and what we have shown is that in a sense this is the tightest lower bound on $< P_e >$ that exists for C not too much greater than $H(U)$.

For the second example about calculating $R(D^*)$, suppose the source space U and the destination space V are the set of real numbers and suppose that the distortion measure is a difference distortion measure, i.e. a measure satisfying

$$D(u,v) = D'(u-v) \tag{3.71}$$

for all u, v and some function D'. Then the constraint set H becomes

$$H = \left\{ h(u) : \int dQ_u \, h(u) \exp\left[-\varrho D'(u-v)\right] \leq 1, \text{ all } v \right\}. \tag{3.72}$$

Let μ be the ordinary Lebesgue measure on the real line and let $f(\varrho)$ be

$$f(\varrho) = \left[\int_u \exp\left[-\varrho D'(u)\right] d\mu \right]^{-1} \tag{3.73}$$

Let

$$h(u) = f(\varrho) \frac{d\mu}{dQ_u} \tag{3.74}$$

over the region where μ is absolutely continuous with respect to Q_u. We see that this $h(u)$ is in H and thus

(3.75) $$R_0(\varrho) \geq \ln f(\varrho) + \int dQ_u \ln \frac{d\mu}{dQ_u}$$

For the case of a square distortion measure (i.e., $D(u,v) = (u,v)^2$), we can perform the integration in (3.73), obtaining $f(\varrho) = \sqrt{\varrho/\pi}$. If we further assume that Q_u is a Gaussian (normal) distribution with zero mean and variance A then (3.75) simplifies to

(3.76) $$R_0(\varrho) \geq \frac{1}{2} \ln (2 \varrho e A)$$

Using $(\varrho) = 1/(2D^*)$ for the tightest bound on $R(D^*)$, we obtain

(3.77) $$R(D^*) \geq \begin{cases} \frac{1}{2} \ln (A/D^*) \, ; & D^* \leq A \\ 0 & ; \; D^* > A \end{cases}$$

Furthermore, if we let ω be a Gaussian distribution with zero mean and variance $A - \dfrac{1}{2\varrho}$, we find that (3.76) and hence (3.77) are satisfied with equality.

In the final part of this section we shall prove the coding theorem for memoryless discrete time sources with a distortion measure and the converse theorem for arbitrary discrete time sources. We first do the converse, since in principle it is trivial. Assume that the output of the source is processed, then transmitted over a channel of capacity C nats per source symbol. The channel output is then processed, delivering a sequence of symbols to the destination. The converse to the coding theorem states that for any D^*, if $C < R(D^*)$, then the average distortion must be greater than D^*. To see this, we observe that if $C < R(D^*)$, then we can pick n large enough that $C < R_n(D^*)$. By the definition of $R_n(D^*)$, the average mutual information per letter between the sequence $u_1, \ldots, u_n$ and $v_1, \ldots, v_n$ must be at least $R_n(D^*)$ to achieve average distortion D^* or less. By the data processing theorem, however, the average mutual information over the channel must be at least as large as that between source and destination, and thus must be at least $R_n(D^*)$. This average information per letter, however, can be no greater than C by the definition of channel capacity, so that it cannot be at least $R(D^*)$. Thus, the average distortion must exceed D^*.

We have not stated this as a formal theorem because of the difficulties inherent in defining capacity as a maximum average mutual information for arbitrary channels with memory. This difficulty should be clear after our study of finite state channels. One can state converses to the source coding theorem for each class of channels (see Gallager, section 9.2), but the point is that these theorems are really theorems about how to define capacity for strange channels. The important thing here is to understand that the coding theorem converse for sources with a distortion measure is an obvious consequence of the definition of $R(D^*)$, the data processing theorem, and the various definitions of capacity.

Before proving the coding theorem itself for sources with a distortion measure, we must introduce the idea of a block code. A source code of length L with M code words is a set of M code words, $\vec{v}_1, \ldots, \vec{v}_M$, where each code word $\vec{v}_M$ is a sequence of L destination symbols. The encoding rule for such a code is to map each source sequence for length L, $\vec{u} = (u_1, \ldots, u_L)$ into that code word $\vec{v}_m$ for which $D_L(\vec{u}, \vec{v}_m)$ is minimum. If there are several m for which $D_L(\vec{u}, \vec{v}_m)$ is minimum, it makes no difference which one is chosen, but to be specific, we assume the smallest m for which $D_L(\vec{u}, \vec{v}_m)$ is minimum is chosen. We denote the $\vec{v}_m$ into which u is encoded by $\vec{v}(\vec{u})$.

The way such codes can be used is as follows : each sequence of L source letters $\vec{u}$ is first encoded into $\vec{v}(\vec{u})$. The M code words are then mapped into binary sequences. The resulting sequence of binary digits is then encoded for transmission over a noisy channel. In the absence of decoding errors, the code words $\vec{v}_m$ can be retrieved, and the average distortion per letter is $E\left[D_L(u, \vec{v}(\vec{u}))\right]$.

Theorem (Coding Theorem) : Let $R(D^*)$ be the rate distortion function of a discrete time meoryless source and assume that a finite set of destination letters $v_1, \ldots, v_J$ exist such that

$$E\left[\min_{1 \le j \le J} D(u, v_j)\right] < \infty \tag{3.78}$$

when the expectation is over the source. For any $D^* > 0$, any $\delta > 0$, and all sufficiently large L, there exists a source code of length L where the entropy of the set of code words satisfies

$$H(V^L) \le L\left[R(D^*) + \delta\right] \tag{3.79}$$

and the average distortion per letter satisfies

$$E\left[D_L(\vec{u}, \vec{v}(\vec{u}))\right] \le D^* + \delta . \tag{3.80}$$

Furthermore, if $\ln J \le R(D^*)$, the number of code words, M_1 also satisfies

$$\ln M_1 \le L\left[R(D^*) + \delta\right]. \tag{3.81}$$

Discussion : The restriction in (3.78) is not very restrictive, and most sources of interest satisfy this restriction with $J = 1$, in which case (3.81) is always valid. For the cases in which (3.78) is not satisfied, it is easy to see that the average distortion of all source codes is infinite. We also observe that if (3.79) is satisfied, then the code words can be mapped by a prefix condition code into arbitrarily little more than $(R(D^*) + \delta) \ln 2$ binary digits per source digit. Given a channel whose capacity exceeds $R(D^*) + \delta$, we know that these digits can be transmitted with arbitrarily small error probability, and thus the average distortion will be at most $D^* + \delta$. We can say little about the distortion that arises from channel decoding errors, except that the probability of such events can be made arbitrarily small.

Before proving the theorem, we state and prove a fundamental lemma.

Lemma: For a given source, distortion measure, and block length L, let P be a probability measure on sequences of L source letters and L distination letters for which the marginal distribution on the source sequence is the given source distribution Q. Consider an ensemble of source codes of length L with M code words each in which each code word is chosen independently by the marginal distribution of P on the destination sequences. Let $P_c(D > \hat{D})$ be the probability, over the ensemble of codes and over the source distribution that $D(\vec{u}, \vec{v}(\vec{u}))$ exceeds $\hat{D}$. Then

$$P_c(D > \hat{D}) \leq P(A) + \exp\left[-Me^{-L\hat{R}}\right] \tag{3.82}$$

when $\hat{R}$ is an arbitrary number and

$$A = \left\{\vec{u}, \vec{v} : I(\vec{u}; \vec{v}) > L\hat{R} \text{ or } D(\vec{u}, \vec{v}) > \hat{D}\right\} \tag{3.83}$$

Now for any given $\vec{u}$, $D(\vec{u}, \vec{v}(\vec{u}))$ cannot exceed $\hat{D}$ unless $D(\vec{u}, \vec{v}_m) > \hat{D}$ for each m. It follows that $D(\vec{u}, \vec{v}(\vec{u}))$ cannot exceed $\hat{D}$ unless $(\vec{u}, \vec{v}_m) \in A$ for each m. Let

$A_{\vec{u}} = \left\{ \vec{v} : \vec{u}, \vec{v} \in A \right\}$. Then, for the given $\vec{u}$, $\omega(A_{\vec{u}})$ is the probability over the ensemble of codes that $\vec{v}_m \in A_u$ for any particular m. Since the code words are independently chosen, $\left[\omega(A_{\vec{u}})\right]^M$ is the probability that $\vec{v}_m \in A_u$ for all m, $1 < m < M$. Then

$$P_c(D > \hat{D}) \leq \int_{\vec{u}} \left[\omega(A_{\vec{u}})\right]^M dQ$$

Letting $A^c_{\vec{u}}$ be the complement of the set $A_{\vec{u}}$, this becomes

$$P_c(D > \hat{D}) \leq \int_{\vec{u}} \left[1 - \omega(A^c_{\vec{u}})\right]^M dQ \tag{3.84}$$

For $v \in A^c_{\vec{u}}$, we see from (3.83) that

$$\ln\left[\frac{dP}{d(Q x \omega)}(\vec{u}, \vec{v})\right] \leq L\hat{R}$$

$$\frac{dP}{d(Q x \omega)}(\vec{u}, \vec{v})\, e^{-L\hat{R}} \leq 1$$

$$\omega(A^c_u) = \int_{A^c_u} d\omega \geq e^{-L\hat{R}} \int_{A^c_u} \frac{dP}{d(Q x \omega)}(\vec{u}, \vec{v})\, d\omega$$

The integral is equal (almost everywhere with respect to Q) to the conditional probability of A^c_u given u with the probability measure P.

$$\omega(A^c_u) \geq e^{-L\hat{R}} p(A^c_u | \vec{u}) \tag{3.85}$$

Next we need the inequality

$$\left[1-\alpha x\right]^{M} = e^{M\ln(1-\alpha x)} \le e^{-M\alpha x}$$

(3.86) $$\le 1-x+e^{-M\alpha}\ ; \quad \alpha \ge 0\ ,\ 0 \le x \le 1$$

The final inequality is verified by noting that equality holds for $x=0$ and $x=1$ and that $e^{-M\alpha x}$ is convex U in x. Letting $x = P\left(A_u^c \mid \vec{u}\right)$ and $\alpha = e^{-L\hat{R}}$ and substituting (3.85) into (3.84), this becomes

$$P_c\left(D > \hat{D}\right) \le \int \left\{1 - P\left(A_u^c \mid u\right) + \exp\left[-Me^{-L\hat{R}}\right]\right\} dQ .$$

Integrating, we get (3.82), completing the proof of the lemma.

Intuitively, the lemma is used as follows : for a given D^*, we chose P to yield an average distortion D^* and an average mutual information $R\left(D^*\right)$. We then choose $\hat{D}$ slightly larger than D^* and $\hat{R}$ slightly larger than $R\left(D^*\right)$. The law of large numbers then makes $P\left(A\right)$ small for large L and choosing M slightly larger than $\exp L\hat{R}$ makes the final term in (3.82) small. Thus, over the ensemble of codes, the distortion is rarely larger than $\hat{D}$, which is close to D^*.

<u>Proof of Theorem</u> : For given $D^* > 0$ and $\delta > 0$, let P_1 be a joint measure over U, V for which $E\left[D\left(u, v\right)\right] \le D^*$ and $I\left(U ; V\right) \le R\left(D^*\right) + \delta/4$. For a given L to be chosen later, we apply the lemma where P is the Lth order product measure

of P_1, $\hat{D} = D^* + \delta/2$, $\hat{R} = R(D^*) + \delta/2$, and

$$M = \left\lfloor \exp\left\{ L\left[R(D^*) + 3\delta/4\right]\right\}\right\rfloor \tag{3.87}$$

where $\lfloor x \rfloor$ means the integer part of x. Then (3.82) becomes

$$P_c(D > \hat{D}) \leq P(A) + \exp\left\{-e^{L\delta/4} + 1\right\}. \tag{3.88}$$

Since $\frac{1}{L} I(\vec{u}; \vec{v})$, according to the probability measure P, is the sample mean of L independent identically distributed random variables of mean at most $R(D^*) + \delta/4$ and $D(u, v)$ is the sample mean of L independent, identically distributed random variables of mean at most D^*, the weak law of large numbers asserts that $P(A)$ approaches 0 with increasing L.

Now for each L pick a specific code from the ensemble for which the probability that the average distortion exceeds D satisfies (3.88). For this code let

$$B = \left\{ \vec{u} : D(\vec{u}, \vec{v}(\vec{u})) \geq D^* + \delta/2 \right\} \tag{3.89}$$

We next add to this code an additional set of J^L code words, one for each sequence of letters from the set $v_1, \ldots, v_J$ in (3.78). Our encoding rule is to map each $\vec{u} \notin B$ into the minimum distortion $\vec{v}_m$ in the original code and to map each $\vec{u} \in B$ into the best word from the new set. The entropy of the entire set of code words is upper bounded by

$$H(V^L) \le H(V^L | u \notin B)[1 - Q_L(B)]$$
$$+ H(V^L | u \in B) Q_L(B) + H(Q_L(B))$$

(3.90) $$\le L\left[R(D^*) + \frac{3\delta}{4} + Q_L(B)\ln J + \frac{\ln 2}{L}\right]$$

Since $Q_L(B)$ approaches zero with increasing L, $H(V^L) \le$ $\le L[R(D^*) + \delta]$ for all sufficiently large L. Also, if $\ln J \le$ $\le R(D^*)$, the total number of code words is

$$M_1 = M + J^L \le 2 \exp\left\{L\left[R(D^*) + \frac{3\delta}{4}\right]\right.$$

(3.91) $$\le \exp \; L[R(D^*) + \delta]$$

for all sufficiently large L.

Next we upper bound the average distortion. Let

(3.92) $$z(u_\ell) = \min_{1 \le i \le J} D(u_\ell, v_j)$$

Then, for $\vec{u} \in B$, the distortion will be $\frac{1}{L} \sum_{\ell=1}^{L} z(u_\ell)$.
For $u \notin B$, the distortion is at most $D^* + \delta/2$. Thus the average distortion for the code, $\bar{D}$, satisfies

(3.93) $$\bar{D} \le D^* + \delta/2 + \int_{\vec{u} \in B} \frac{1}{L} \sum_{\ell=1}^{L} z(u_\ell)\, dQ_L$$

(3.94) $$= D^* + \delta/2 + \frac{1}{L} \sum_{\ell=1}^{L} \int_{u_\ell} z(u_\ell)\, Q_L(B | u_\ell)\, dQ_1$$

Now let z' be a number to be determined later and let $B_1 = \{u : z(u) > z'\}$. Then

$$\bar{D} \leq D^* + \frac{\delta}{2} + \frac{1}{L} \sum_{\ell=1}^{L} \left[\int_{u_\ell \notin B_1} z' Q_L(B|u_\ell) dQ_1 + \int_{u_\ell \in B_1} z(u_\ell) dQ_1 \right]$$

$$\leq D^* + \frac{\delta}{2} + z' Q_L(B) + \int_{z=z'}^{\infty} z \, dF(z) \qquad (3.95)$$

when $F(z)$ is the distribution function of $z(u)$. Since by assumption $E(z) < \infty$, we can choose z' large enough that

$$\int_{z=z'}^{\infty} z \, dF(z) < \delta/4$$

Then for all L sufficiently large, $z' Q_L(B) \leq \delta/4$, and thus $\bar{D} \leq D^* + \delta$ completing the proof.

Random Trees and Tree Codes

A random tree is a probabilistic system much like a random walk. In a random walk, a particle moves up or down as time progresses in accordance with some stochastic law. A random tree, on the other hand, starts with one particle at time zero, this particle branches into a number of particles, each of which move up or down in accordance with a stochastic law, and at each successive time unit, each existing particle again branches into a number of descendant particles, each of which move up or down according to a stochastic law.

We shall first give a precise mathematical description of random trees. Then we shall develop two sets of results about random trees, one stemming from random walk type analysis techniques and the other stemming from branching process analysis techniques. It appears that this work should also be applicable in many other areas, but so far no specific applications are known, except in the area of sequential decoding.

As with Markov chains and random walks, it is somewhat simpler and more natural to describe a random tree in terms of its evolution in time rather than in terms of the underlying sample space. A sample function of the initial part of a random tree is shown in Figure 4.1, with time plotted on the horizontal axis and particle levels plotted on the vertical axis. The tree starts at time 0 with one particle at some prespecified level, x. This particle branches into several particles at time 1. This number of particles, M_1, may be a random variable or it may be fixed. We denote the level of the jth of these particles $(1 \leq j \leq M_1)$ by $y_j(1,0,x)$. The reason for the arguments 1 and 0 will become clear presently. In the most general case, $M_1, y_1(1,0,x), y_2(1,0,x), \ldots, y_{M_1}(1,0,x)$ are dependent random variables with some known joint probability distribution. In the case of most interest, however, conditional on M_1, $y_1, y_2, \ldots, y_{M_1}$ are independent and identically distributed. For each j, $1 \leq j \leq M_1$, the jth particle at time 1 branches into a number of particles $M_2(y_j)$ at time 2. The level of the kth particle $(1 \leq k \leq M_2(y_j))$ at time 2 that branches from the jth particle at time 1 is denoted $y_{k,j}(2, k, y_j)$. In the most general case $M_2(y_j)$, $y_{1,j}(2,1,y_j)$, $\ldots$, $y_{M_2(y_j),j}(2,1,y_j)$ are dependent random variables with a joint distribution that depends on y_j but not on j. We assume, however, that the above set of random variables (conditional on y_j) is statistically independent of all the other random variables so far defined. In more heuristic terms, the branching of a particle and the motion of its descendents is independent of the branching and motion of all the other particles.

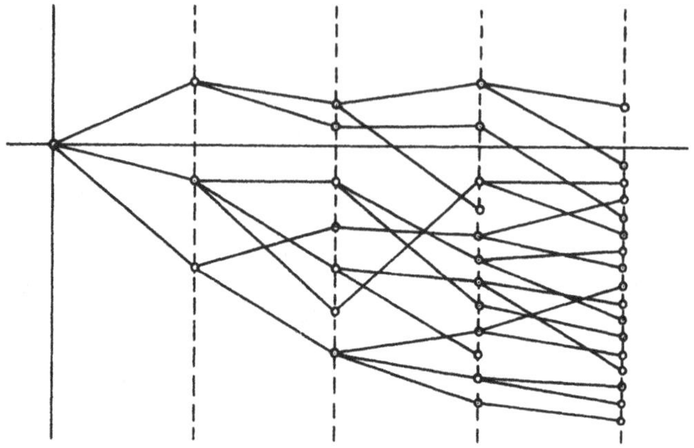

Sample Function of the Initial Part of a Random Tree
Figure 4. 1.

The total number of particles at time 2 is given by

(4. 1) $$M_2 = \sum_{j=1}^{M_1} M_2(y_j)$$

It is often convenient notationally to suppress the information about the parent of a particle at time 2 and to simply order these particles from 1 to M_2, in which case we denote the level of the jth particle at time 2 by $y_j(2,0,x)$ $(1 \le j \le M_2)$. Similarly at any time n, $(n \geq 1)$, we can order the particles from 1 to M_n, where M_n is the total number of particles at time n. We denote the level of the jth particle at time n by $y_j(n,0,x)$. Let $M_n(y_j)$ be the number of particles at time n that the jth particle at time n-1 branches into. Similarly let $y_{k,j}(n,n-1,y_j)$ be the level of the kth descendent $1 \le k \le M_n(y_j)$ at time n

of the jth particle at time n-1. We assume as before that the set of random variables $M_n(y_j)$, $y_{1,j}(n, n-1, y_j)$, ..., $y_{M_n(y_j),j}(n,n-1,y_j)$ are statistically dependent with a joint distribution function which depends on the level y_j but not on j. Also this set of random variables is statistically independent (conditional on y_j) of all other particle levels and numbers of descendants at times less than or equal to n. In other words, we can think of generating a random tree by taking a random tree generated out to time n and performing an independent experiment on each particle in the tree at time n. These experiments will depend on n and on the level of the particle but nothing else, and the outcome of each experiment will be the number of descendants at time $n+1$ of the given particle and the level of each of those descendants. By starting at $n = 0$ and performing these experiments successively for each n, the tree is stochastically defined, in terms of these experiments, for each time n.

Along with the level of each particle, it is useful to refer to the increments between the level of one particle and the levels of its descendants at the next time unit. We denote these increments by

$$z_{k,j}(n, n-1, y_j) = y_{k,j}(n, n-1, y_j) - y_j(n-1, 0, x) \tag{4.2}$$

Clearly, specifying the joint distribution of $M_n(y_j)$ and $\{z_{k,j}(n, n-1, y_j)\}$ $(1 \leq k \leq M_n(y_j))$ is equivalent to specifying the joint distribution of $M_n(y_j)$ and $\{y_{k,j}(n, n-1, y_j)\}$

$(1 \leq k \leq M_n(y_j))$. If this joint distribution of $M_n(y_j)$ and $\{z_{k,j}(n, n-1, y_j)\}$ is not a function of n, we say that the random tree is stationary in time. If it is not a function of y_j, we say that the random tree is stationary in level. We call a tree which is stationary in both time and level a completely stationary random tree.

For a completely stationary random tree, the total number of particles at each time,

$$M_n = \sum_{j=1}^{M_{n-1}} M_n(y_j),$$

forms a branching process since under these circumstances the random variables $M_n(y_j)$ are independent and identically distributed. Also if a completely stationary random tree satisfies $M_n = 1$ for all n, then the random tree reduces to a random walk. It is tempting to think that a completely stationary random tree with $M_n \neq 1$ can be thought of as a set of independent random walks, but this is false since the levels of two particles which branch from the same particle will be statistically coupled by the level of the parent particle.

We shall be particularly interested in completely stationary random trees for which the expected number of particles grows with n but where the particles drift downward in level. If the drift is sufficiently great relative to the growth rate, then for large n all the particles will with high probability be at very low levels. In order to study this type of problem, it is convenient to impose absorbing barriers on the random tree. In random walk theorem, one generally terminates the random walk when the particle crosses an absorbing barrier. Here, since there are generally many particles, each doing their own thing, we cannot terminate the tree. It is more convenient to consider particles that have crossed the barrier as being "frozen". Such particles remain for all future time without further branching or level changes. We consider such "freezing barriers" as being placed at levels say $+\alpha$ and $-\beta\,(\alpha, \beta > 0)$ with the random tree starting at x, $-\beta < x < \alpha$. Then in terms of our previous notation, we take $M_n(y_j) = 1$ with probability 1 for $y_j > \alpha$ and for $y_j < -\beta$, and we take $z_{i,j}(n, n-1, y_j) = 0$. Figure 4.1 has been redrawn with freezing barriers in Figure 4.2.

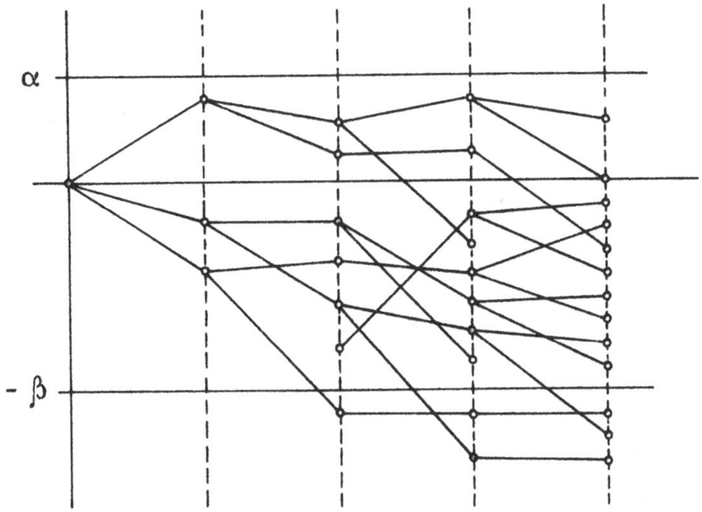

Random Tree with Freezing Barriers
Figure 4.2

It should be observed that there are fewer particles between the barriers in Figure 4. 2 than in Figure 4. 1. In particular only the particles remain for which the path to the origin stays between the barriers. We now proceed to analyze random trees by two different techniques.

Random Walk Techniques

Assume a completely stationary random tree* with freezing barriers at α and $-\beta\,(\alpha, \beta > 0)$. We can assume without loss of generality that the starting point is at $x = 0$. For the jth particle at time n, define

(4. 3)
$$\ell(j) = \begin{cases} n & \text{if } \; y_j(n,0,0) \in (-\beta, \alpha) \\ m & \text{if } \; y_j(n,0,0) \notin (-\beta, \alpha) \text{ where } m \text{ is the first time at which the path from the origin to } y_j \text{ crossed the threshold.} \end{cases}$$

Then define the quantity

(4. 4)
$$T_n(s,r) = \sum_{j=1}^{M_n} e^{s y_j(n,0,0) + r\ell(j)}$$

(*) A similar analysis can be carried out without the stationarity in time assumption. It is somewhat messier, however, and not of great general interest.

Using (4. 2) and (4. 3), we see that

$$T_n(s,r) = \sum_{j=1}^{M_{n-1}} e^{sy_j(n-1,0,0) + r\ell(j)} f_j(y_j, s, r) \qquad (4.5)$$

$$f_j(y_j, s, r) = \begin{cases} e^r \sum_{k=1}^{M_n(y_j)} \exp\left[s z_{k_j}(n, n-1, y_j); \right. & y_j \in (-\beta, \alpha) \\ 1 & ; \ y_j \notin (-\beta, \alpha) \end{cases}$$

(4. 6)

For $n=1$, we take $y_j(n-1, 0, 0)$ as zero.

The distribution function of $f_j(y_j, s, r)$ depends only on whether $y_j \in (-\beta, \alpha)$ or not. It does not depend on n, j or the value of y_j within $(-\beta, \alpha)$. For $y_j \in (-\beta,\alpha)$, $f_j(y_j,s,r)$ has the same distribution function as

$$f_1(0, s, r) = e^r \sum_{k=1}^{M_1} \exp\left[s z_{k,1}(1,0,0) \right] \qquad (4.7)$$

which corresponds to the increments from time 0 to time 1. Suppose now that values of s, r exist such that $E\left[f_1(0,s,r)\right] = 1$. With such values for s, r, we see from (4. 6) that $E\left[f_j(s,r)\right]=1$ for all values of y_j. We can then take the expected value of $T_n(s,r)$ in (4. 5), first taking the expected value of each $f_j(y_j, s, r)$ for a given set of values y_j at time $n-1$, and then taking the expected value over the set of values y_j. We obtain therefore

$$E\left[T_n(s,r)\right] = E\left[T_{n-1}(s,r)\right] \tag{4.8}$$

Since $T_0(s,r) = 1$, we see by induction that

$$E\left[T_n(s,r)\right] = 1 \ ; \ \text{all } n \geq 0 \tag{4.9}$$

We are not particularly interested in the most general conditions for which $E\left[f_1(0,s,r)\right] = 1$. However, conditional on any given value of M_1, we observe that $E\left[f_1(0, s,r) \mid M_1\right]$ is just e^r times a sum of moment generating functions. If the positive and negative tails of each random variable $z_{k,1}$, conditional on each M_1, are bounded by some fixed decaying exponential, and if $E(M_1) < \infty$, then $E\left[f_1(0,s,r)\right]$ will exist over a range of s including the origin and will be convex $\cup$ in s over this range. If the random variables $z_{k,1}$ are bounded and take on both positive and negative values, then $E\left[f_1(0,s,r)\right]$ will exist for all s, r and will approach ∞ as s approaches $+\infty$ and $-\infty$. As can be seen from Figure 4. 3, there is some maximum r, r_{max}, for which an s exists such that $E\left[f_1(0,s,r)\right] = 1$. For each $r < r_{max}$, there are two values of s, say $s_1(r) < s_2(r)$ such that $E\left[f_1(0,s,r)\right] = 1$.

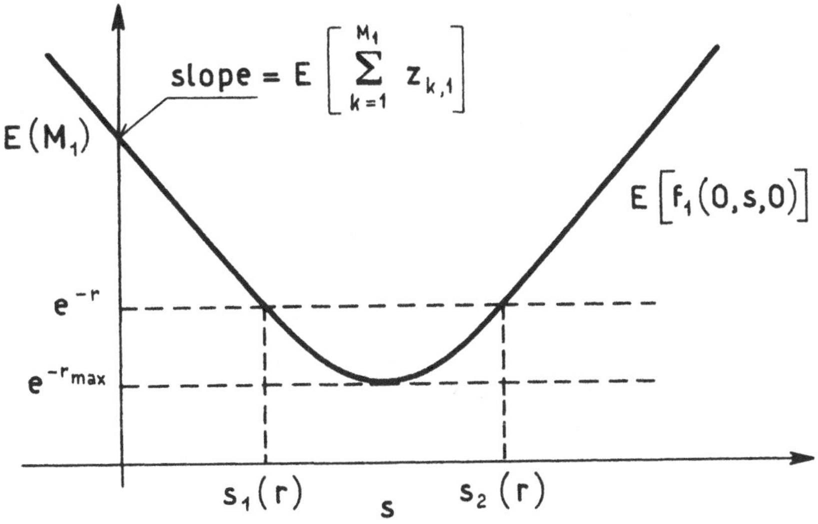

Behavior of $E[f_1(0,s,r)] = e^r E[f_1(0,s,0)]$ where $E(M_1) < \infty$ and each $z_{k,1}(1,0,0)$ for $1 \le k \le M$, is bounded.

Figure 4. 3

Now suppose the random variables $z_{k,1}(1,0,0)$ are bounded above by δ^+ and bounded below by $-\delta^-$. Then, a particle crossing the α barrier has a value between α and $\alpha+\delta^+$ and a particle crossing the barrier at $-\beta$ has a value between $-\beta$ and $-\beta - \delta^-$. Assume that for a particular $r, r<r_{max}, s_1(r)$ and $s_2(r)$ are both positive (the modification for negative $s_1(r)$ or negative $s_1(r)$ and $s_2(r)$ should be obvious).

Let $N_{\ell,\alpha}$ be the number of particles crossing the α barrier at time ℓ and $N_{\ell,\beta}$ be the number of particles crossing the β barrier at time ℓ. Then for $i = 1$ or 2

$$T_n(s_i(r), r) \geq \sum_{\ell=1}^{n} \left[N_{\ell,\alpha}\, e^{s_i(r)\alpha + r\ell} + N_{\ell,\beta}\, e^{-s_i(r)[\beta+\delta^-]+r\ell} \right] +$$

$$+ R(n, i, r) \tag{4.10}$$

$$T_n(s_i(r), r) \leq \sum_{\ell=1}^{n} \left[N_{\ell,\alpha}\, e^{s_i(r)[\alpha+\delta^+]+r\ell} + N_{\ell,\beta}\, e^{-s_i(r)\beta + r\ell} \right] +$$

$$+ R(n, i, r) \tag{4.11}$$

where

$$R(n,i,r) = \sum_{j\,:\,y_j \in (-\beta,\alpha)} e^{s_i(r)\, y_j(n,0,0) + rn} \tag{4.12}$$

We now show that the expected value of $R(n,i,r)$ goes to O with increasing n for any fixed $r < r_{max}$. Since $y_j(n,0,0) \leq \alpha$ for all j in the sum,

$$E\left[R(n,i,r)\right] \leq e^{s_i(r)\alpha + rn}\, E\left[N_n\right] \tag{4.13}$$

when N_n is the number of j such that $y_j(n,0,0) \in (-\beta,\alpha)$. On the other hand, using r_{max} in place of r,

$$1 = E\left[T_n(s_i(r_{max}), r_{max})\right] \geq E\left[R(n,i,r_{max})\right]$$

$$\geq e^{-s_i(r_{max})\beta + r_{max}n} \; E\left[N_n\right]. \tag{4.14}$$

Note from Figure 4. 3 that $s_1(r_{max}) = s_2(r_{max})$ so the i is unnecessary. Using (4. 14) as an upper bound on $E\left[N_n\right]$ and substituting this in (4. 13),

$$E\left[R\,(n, i, r)\right] \leq e^{s_i(r)\alpha + s_i(r_{max})\beta - n(r_{max} - r)} \tag{4.15}$$

This approaches 0 as n approaches ∞. Thus we can take the expected value of (4. 10) and (4. 11), and pass to the limit $n \to \infty$ obtaining, for $r < r_{max}$

$$1 = E\left[T_\infty(s_i(r), r)\right] \begin{cases} \geq A(r)e^{s_i(r)\alpha} + B(r)e^{-s_i(r)(\beta + \delta^-)} \\ \leq A(r)e^{s_i(r)(\alpha + \delta^+)} + B(r)e^{-s_i(r)\beta} \end{cases} \tag{4.16}$$

where

$$A(r) = \sum_{\ell=1}^{\infty} E\left[N_{\ell,\alpha}\right] e^{r\ell} \tag{4.17}$$

$$B(r) = \sum_{\ell=1}^{\infty} E\left[N_{\ell,\beta}\right] e^{r\ell}. \tag{4.18}$$

We can now use the upper bound in (4. 16) for $i = 1$ and the lower bound for $i = 2$ to obtain an upper bound on $A(r)$ and a lower bound on $B(r)$. Using the opposite inequalities, we lower bound $A(r)$ and upper bound $B(r)$. The results are

$$(4.19)\qquad A(r)\begin{cases} \leq e^{-s_2(r)\alpha}\left\{\dfrac{1-\exp\left[-\beta(s_2(r)-s_1(r))-s_2(r)\delta^-\right]}{1-\exp\left[-(\alpha+\beta)(s_2(r)-s_1(r))+s_1(r)\delta^+-s_2(r)\delta^-\right]}\right\} \\[2ex] \geq e^{-s_2(r)(\alpha+\delta^+)}\left\{\dfrac{1-\exp\left[-\beta(s_2(r)-s_1(r))+s_1(r)\delta^-\right]}{1-\exp\left[-(\alpha+\beta)(s_2(r)-s_1(r))+s_1(r)\delta^--s_2(r)\delta^+\right]}\right\} \end{cases}$$

$$(4.20)\qquad B(r)\begin{cases} \leq e^{s_1(r)(\beta+\delta^-)}\left\{\dfrac{1-\exp\left[-\alpha(s_2(r)-s_1(r))-s_2(r)\delta^+\right]}{1-\exp\left[-(\alpha+\beta)(s_2(r)-s_1(r))-s_2(r)\delta^++s_1(r)\delta^-\right]}\right\} \\[2ex] \geq e^{s_1(r)\beta}\left\{\dfrac{1-\exp\left[-\alpha(s_2(r)-s_1(r))+s_1(r)\delta^+\right]}{1-\exp\left[-(\alpha+\beta)(s_2(r)-s_1(r))-s_2(r)\delta^-+s_1(r)\delta^+\right]}\right\} \end{cases}$$

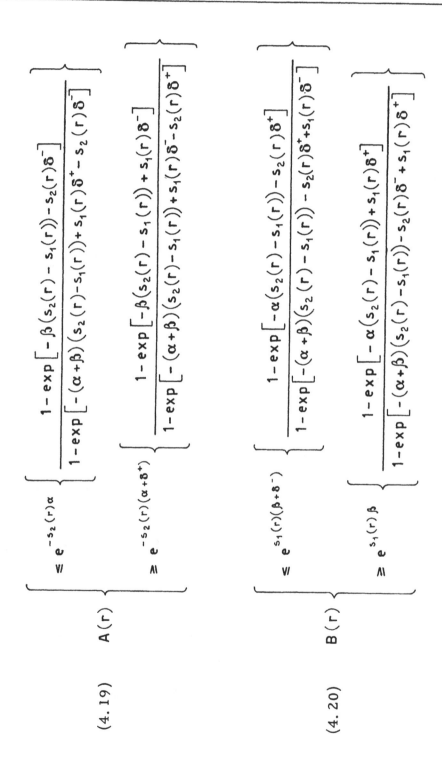

All of the expressions in braces above are less than 1 and approach 1 as α and β increase. If the maximum increments δ^+ and δ^- are also negligible with respect to α and β, then we have $A(r) \approx \exp[-s_2(r)\alpha]$ and $B(r) \approx \exp[s_1(r)\beta]$. These final approximations are valid independent of the sign of $s_1(r)$ and $s_2(r)$, although the exact bounds must be modified somewhat if s_1 and s_2 are not positive.

In many situations it is convenient to have similar expressions valid for $r = r_{max}$. In that case the bounds in (4. 19) and (4. 20) break down, but we can work directly from (4. 10) to obtain

$$A(r_{max}) \le e^{-s(r_{max})\alpha} \tag{4. 21}$$

$$B(r_{max}) \le e^{s(r_{max})(\beta+\delta^-)} \tag{4. 22}$$

It should be clear from the results so far that the character of a completely stationary random tree depends critically on whether r_{max} is greater or less than 0. First assume $r_{max} \ge 0$. The expected total number of particles that cross the α threshold is then $A(0)$ which is bounded by (4. 19) or by (4. 21) if $r_{max} = 0$. This expected number is upper bounded by the expression $\exp[-s_2(0)\alpha]$ which is exponentially decreasing in α (assuming as before that $s_2(0) > 0$). Clearly $\exp[-s(0)\alpha]$ is also an upper bound on the probability that one or more particles will cross the α threshold. Likewise $B(0)$, bounded by (4. 20) or by (4. 22), is the expected number of particles that cross the $-\beta$ threshold. This number is exponentially increasing in β (for $s_1(0) > 0$), which is to be expected since $s_1(0) > 0$ corresponds to the case where the expected number of particles in the unrestricted tree is increasing and they are drifting downward.

We can also obtain upper bounds on the expected number of α or β crossings before or after some given time n. For $r > 0$ (assuming $r_{max} > 0$),

$$e^{-s_2(r)\alpha} \geq A(r) = \sum_{\ell} E(N_{\ell,\alpha}) e^{r\ell}$$

$$\text{(4.23)} \qquad \geq \sum_{\ell \geq n} E(N_{\ell,\alpha}) e^{r\ell} \geq e^{rn} \sum_{\ell \geq n} E(N_{\ell,\alpha}).$$

Thus, the expected number of α crossing at times n or greater is

$$\text{(4.24)} \qquad \sum_{\ell \geq n} E(N_{\ell,\alpha}) \leq e^{-s_2(r)\alpha - rn} \quad ; \qquad r \geq 0$$

This bound can be optimized by choosing r to satisfy

$$\text{(4.25)} \qquad \frac{n}{\alpha} = -\frac{ds_2(r)}{dr} = \left. \frac{E\left[f_1(0,s,0)\right]}{\frac{\partial}{\partial s} E\left[f_1(0,s,0)\right]} \right|_{s=s_2(r)}$$

The r that satisfies (4.25) is increasing with n. The bound is only useful for $n > -\alpha \left. \frac{ds_2(r)}{dr} \right|_{r=0}$. For smaller values of n, we can take $r < 0$ and obtain

$$\sum_{\ell \leq n} E(N_{\ell,\alpha}) \leq e^{-s_2(r)a - rn} \quad ; \quad r < 0 \tag{4.26}$$

Again the best choice of r for this bound is found from the solution of (4.25). What these bounds indicate is that if crossings of the barrier occur, they are likely to occur at values of n close to $-\alpha \left.\frac{ds_2(r)}{dr}\right|_{r=0}$.

The analogous results for the barrier at $-\beta$ are

$$\sum_{\ell \geq n} E\left[N_{\ell,\beta}\right] \leq \exp\left[s_1(r)\beta - rn\right] \quad ; \quad r \geq 0 \tag{4.27}$$

For $n \geq \beta \left.\frac{ds_1(r)}{dr}\right|_{r=0}$, the tightest bound is achieved by choosing r to satisfy

$$\frac{n}{\beta} = \frac{ds_1(r)}{dr} = - \left.\frac{E\left[f_1(0,s,0)\right]}{\frac{\partial}{\partial s} E\left[f_1(0,s,0)\right]}\right|_{s=s_1(r)} . \tag{4.28}$$

For large n however, it is convenient to simply choose $r = r_{max}$, obtaining

$$\sum_{\ell \geq n} E\left[N_{\ell,\beta}\right] \leq \exp\left[s(r_{max})\beta - r_{max} n\right]. \tag{4.29}$$

For $n < \beta \left.\frac{ds_1(r)}{dr}\right|_{r=0}$, (4.27) is optimized by choosing $r = 0$, but also

$$\sum_{\ell \leq n} E\left[N_{\ell,\beta}\right] \leq \exp\left[s_1(r)\beta - rn\right] \quad ; \quad r < 0 \tag{4.30}$$

when again the tightest bound is achieved by choosing r to satisfy (4. 28). Again we see that most of the contribution to the total expected number of β crossings comes at times close to $\beta \left. \frac{ds_1(r)}{dr} \right|_{r=0}$.

For $r_{max} < 0$, (4. 26) and (4. 30) are still valid, but they simply give bounds on the expected number of barrier crossings out to time n that increase exponentially with n. This does not imply that the expected total number of barrier crossings is infinite because we have no assurance that this is a tight bound. We shall see later that in some cases $r_{max} < 0$ and $\sum_{\ell} E\left[N_{\ell,\beta}\right] < \infty$.

We could go on to obtain other results from this approach. It can be seen that in the special case where $M_n = 1$ for all n (which is the case when the random tree becomes a random walk), then (4. 9) is equivalent to Wald's identity (see, for example, Feller (1966), pp. 568). Thus it is not surprising that many of the things that can be done with Wald's identity can also be done here. The difficulty, however, is that this approach only treats the expected values $E\left[N_{\ell,\beta}\right]$ and $E\left[N_{\ell,\alpha}\right]$. In order to obtain more detailed information about the random variables $N_{\ell,\beta}$ and $N_{\ell,\alpha}$ and also about the particles between the barriers, we must go to a different approach. Not all of the statistical independence assumed in the definition of a random tree has been used here. If one checks the derivation of (4. 8) (from which all else follows) carefully, one sees that we have used the assumption that the random variable $f_j(y_j, s, r)$ in (4.5) is statistically independent of M_{n-1} and of $y_j(n-1, 0, 0)$ (subject to (4.6)). However the branching and increments of two particles either at the same time or at different times and on different paths could be statistically related.

Branching Process Techniques

We first briefly review the elementary theory of branching processes and then develop the kind of generalization required here. Let M_n be the total number of particles at time n and let $x_{j,n}$ be the number of particles that the jth of these M_n particles splits into from time n to $n+1$. Assume that $\{x_{j,n}\}$, $1 \le j \le M_n$, $0 \le n \le \infty$ are all independent and identically distributed with the same distribution as M_1. We have

$$M_{n+1} = \sum_{j=1}^{M_n} x_{j,n} . \qquad (4.31)$$

Equivalently, we can let $M_n^{(j)}$ be the number of progeny at time $n+1$ of the jth particle at time 1 and write

$$M_{n+1} = \sum_{j=1}^{M_1} M_n^{(j)} . \qquad (4.32)$$

Clearly $\{M_n^{(j)}\}$, $1 \le j \le M_1$, are independent with the same distribution as M_n. For each n, let $\mu_n(s)$ be the semi-invariant moment generating function of M_n,

$$\mu_n(s) = \ln E\left[e^{sM_n}\right] . \qquad (4.33)$$

Then

$$\mu_{n+1}(s) = \ln E\left[\exp \sum_{j=1}^{M_1} s M_n^{(j)}\right]. \tag{4.34}$$

Conditional on a fixed value of M_1, we can take the expected value of (4. 34) over $M_n^{(j)}$ for each j. Using the independence of these quantities, we obtain

$$\mu_{n+1}(s) = \ln E_{M_1}\left\{\prod_{j=1}^{M_1} E_{M_n}(j) \exp sM_n^{(j)}\right\} \tag{4.35}$$

where the subscripts on the expectation signs denote the random variables being operated on. Using (4. 33) and the fact that $M_n^{(j)}$ has the same distribution as M_n,

$$\mu_{n+1}(s) = \ln E_{M_1}\left\{\prod_{j=1}^{M_1} \exp M_n(s)\right\} = \ln E\left[e^{\mu_n(s) M_1}\right]$$

(4. 36)

$$\mu_{n+1}(s) = \mu_1(\mu_n(s)).$$

Now $\mu_1(s)$ is a convex $\cup$ nondecreasing function of s satisfying

$$\mu_1(0) = 0 \; ; \quad \mu_1'(0) = E[M]$$

$$\lim_{s \to -\infty} \mu_1(s) = \ln \Pr[M_1 = 0].$$

In Figure 4. 4, we sketch $\mu_1(s)$ for the case where $\mu_1'(0) > 1$ and $\mu_1'(0) < 1$ and show how to find $\mu_n(s)$ graphically.

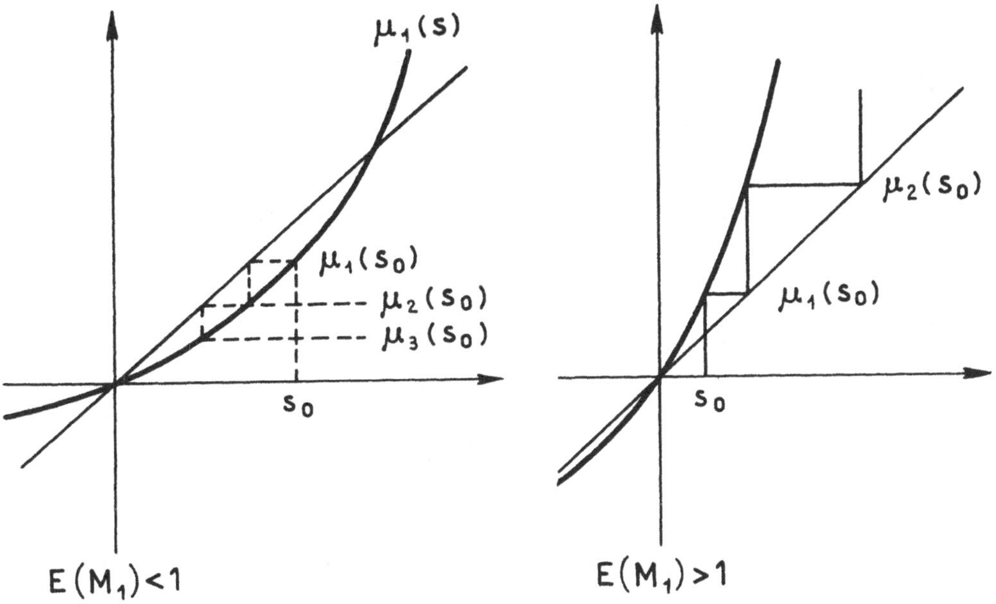

Construction of $\mu_n(s)$

Figure 4.4

It can be seen from the figure that there is something important about the point $s^* \neq 0$ when $s^* = \mu_1(s^*)$. For $E(M_1) < 1$, such a point exists if M_1 takes on some value larger than 1 with non zero probability. For $E(M_1) > 1$, such a point exists if M_1 takes on the value 0 with non-zero probability. We see that if $E(M_1) < 1$, then

$$\lim_{n \to \infty} \mu_n(s) = \begin{cases} 0 & ; \quad s < s^* \\ s^* & ; \quad s = s^* \\ \infty & ; \quad s > s^* \end{cases} \quad . \tag{4.37}$$

For $E(M_1) > 1$

(4. 38)
$$\lim_{n \to \infty} \mu_n(s) = \begin{cases} s^* & ; \quad s < 0 \\ 0 & ; \quad s = 0 \\ \infty & ; \quad s > 0 \end{cases}$$

A number of simple relations are immediately clear.

(4. 39)
$$E\left[M_n\right] = \left\{E\left[M_1\right]\right\}^n$$

(4. 40)
$$\lim_{n \to \infty} Pr\left[M_n = 0\right] = \begin{cases} 1 & ; \quad E\left[M_1\right] < 1 \\ e^{s^*} & ; \quad E\left[M_1\right] > 1 \end{cases}.$$

From (4. 37), it is clear that although the process dies out eventually with probability 1, there are vanishingly small probabilities of extremely large numbers of particles giving rise to $\mu_n(s)$ increasing for $s > s^*$.

The next quantity of interest is the total number of particles out to and including time

(4. 41)
$$N_n = \sum_{\ell=0}^{n} M_\ell .$$

Letting $N_n^{(j)}$ be the total number of particles out to time $n+1$ descending from the jth particle at time 1, we get

(4. 42)
$$N_{n+1} = 1 + \sum_{j=1}^{M_1} N_n^{(j)}$$

Clearly the variables $N_n^{(j)}$ are independent and have the same distribution as N_n. Letting

$$\gamma_{n+1}(s) = \ln E\left[e^{sN_n}\right], \tag{4.43}$$

and going through the same argument as before, we find the recursion formula

$$\begin{aligned} \gamma_{n+1}(s) &= s + \mu_1(\gamma_n(s)) \\ \gamma_0(s) &= s . \end{aligned} \tag{4.44}$$

For $s > 0$, it can be seen by induction that this is an increasing function of n. We can construct $\gamma_n(s)$ graphically from $\mu_1(s)$ much as before as shown in Figure 4.5.

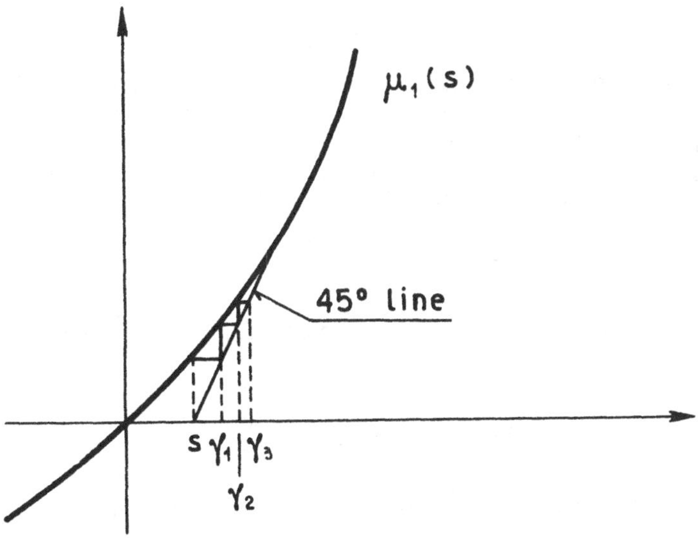

Construction of $\gamma_n(s)$

Figure 4.5

The largest value of s for which $\lim_{n \to \infty} \gamma_n(s)$ exists is found by finding the s_1 for which $\mu'(s_1) = 1$ and erecting a 45° line from s_1, $\mu(s_1)$ to the axis. That intersection is the desired largest s.

Now we are ready to generalize these results to a random tree. The idea is to think of each level on the tree as generating its own branching process, with each of the processes coupled together. If the increment variables are integer valued, we obtain a finite dimensional generalization of a simple branching process, and in general it is an infinite dimensional process. As it turns out, this causes no additional complications.

In conformity with our previous notation, let $y_{k,j}(n, m, y_j)$ be the level of the kth descendant (at time n) of the jth particle at time m. Let $u(\cdot)$ be a real valued function of a real variable, and define

$$\mu_{m,n}(y_j, u(\cdot)) = \ln E\left\{\exp\left[\sum_k u(y_{k,j}(n,m,y_j))\right]\right\} \tag{4.45}$$

$$\mu_{m-1,n}(y_\ell, u(\cdot)) =$$

$$= \ln E \exp\left\{\sum_j \sum_k u \quad y_{k,j}(n, m, y_{j,\ell}(m, m-1, y_j))\right\}. \tag{4.46}$$

First take the expectation over the levels $y_{k,j}$ at time n conditional on a given set of levels at m. This yields

$$\mu_{m-1,n}(y_\ell, u(\cdot)) = \ln E\{\exp \sum_j \mu_{m,n}(y_{j,\ell}(m,m-1,y_\ell), u(\cdot))\}$$

$$= \mu_{m-1,m}(y_\ell, \mu_{m,n}(\cdot, u(\cdot))) \qquad (4.47)$$

Note that $\mu_{m-1,n}$ is a function from $R \times F$ to R where R is the set of real numbers and F is a set of real valued functions of a real variable. μ is not a function of any argument of u but rather of the function itself. For a given function u, $\mu_{m,n}$ on the right side of (4. 47) maps real numbers into real numbers, and thus serves as an appropriate second argument for $\mu_{m-1,m}$.

Now assume that our random tree is stationary in time. Then, $\mu_{m,n}$ depends only on $n-m$, and can be re-written as μ_{n-m}. Then (4. 47) becomes

$$\mu_{n+1}(x, u(\cdot)) = u_1(x, \mu_n(\cdot, u(\cdot))). \qquad (4.48)$$

It is convenient also to have a generating function which deals with the total number of particles out to time n Define

$$\gamma_{m,n}(y_j, u(\cdot)) = \ln E\{\exp \sum_{\ell=m}^{n} \sum_k u(y_{k,j}(\ell,m,y_j))\} \qquad (4.49)$$

where $y_{1,j}(m,m,y_j)$ is y_j. Repeating the argument leading to (4. 47), we get

$$\gamma_{m-1,n}(y_\ell, u(\cdot)) = u(y_\ell) + \mu_{m-1,m}(y_\ell, \gamma_{m,n}(\cdot, u(\cdot))) \qquad (4.50)$$

For a random tree stationary in time this becomes

$$\gamma_{n+1}(x, u(\cdot)) = u(x) + \mu_1(x, \gamma_n(\cdot, u(\cdot))) \tag{4.51}$$

when $\gamma_0(\cdot, u(\cdot)) = u(\cdot)$.

We now develop some properties of these generating functions :

Property 1 : For each $m, n, m < n, \mu_{m,n}(x, u(\cdot))$ and $\gamma_{m,n}(x, u(\cdot))$ are convex $\cup$ in $u(\cdot)$. In other words, for $0 < \lambda < 1$,

$$\mu_{m,n}(x, \lambda u_1(\cdot) + (1-\lambda) u_2(\cdot)) \leq \lambda \mu_{m,n}(x, u_1(\cdot)) + +(1-\lambda)\mu_{m,n}(x, u_2(\cdot)) \tag{4.52}$$

$$\gamma_{m,n}(x, \lambda u_1(\cdot) + (1-\lambda) u_2(\cdot)) \leq \lambda \gamma_{m,n}(x, u_1(\cdot)) + +(1-\lambda)\gamma_{m,n}(x, u_2(\cdot)) \tag{4.53}$$

Equations (4. 52) and (4. 53) follow directly from applying Holder's inequality to (4. 45) and (4. 49).

Property 2 : (Monotone property in u

$$u_1(\cdot) \geq u_2(\cdot) \Longrightarrow \mu_{m,n}(x, u_1(\cdot)) \geq \mu_{m,n}(x, u_2(\cdot)) \tag{4.54}$$

$$u_1(\cdot) \geq u_2(\cdot) \Longrightarrow \gamma_{m,n}(x, u_1(\cdot)) \geq \gamma_{m,n}(x, u_2(\cdot)) \tag{4.55}$$

These again can be seen directly from (4. 45) and (4. 49).

Property 3 :

$$u(\cdot) = 0 \implies \mu_{m,n}(x, u(\cdot)) = \gamma_{m,n}(x, u(\cdot)) = 0 . \quad (4.56)$$

Property 4 :

$$u(\cdot) \geq 0 \implies \mu_{m,n}(x, u(\cdot)) \geq 0 ; \ \gamma_{m,n}(x, u(\cdot)) \geq 0 . \quad (4.57)$$

This follows from properties 2 and 3.

Property 5 :

For a random tree that is stationary in time :

$$\mu_1(x, u(\cdot)) \geq u(x) \implies \mu_{n+1}(x, u(\cdot)) \geq \mu_n(x, u(\cdot)) \quad (4.58)$$

$$\mu_1(x, u(\cdot)) \leq u(x) \implies \mu_{n+1}(x, u(\cdot)) \leq \mu_n(x, u(\cdot)) \quad (4.59)$$

$$u(x) \geq 0 \implies \gamma_{n+1}(x, u(\cdot)) \geq \gamma_n(x, u(\cdot)) \quad (4.60)$$

Equation (4.58) and (4.59) follow by induction from (4.48) and (4.54). Equation (4.60) follows by induction from (4.51) and (4.54).

Property 6 :

Let T be the set of functions $u(\cdot)$ for which

$$\mu_1(x, u(\cdot)) \leq u(x) . \quad (4.61)$$

Then, the set T is convex.

To see this, suppose that u_1 and u_2 are in T

From property 1, for any λ, $0 < \lambda < 1$,

$$\mu_1(x, \lambda u_1 + (1-\lambda) u_2) \leq \lambda \mu_1(x, u_1) + (1-\lambda)\mu_1(x, u_2)$$

$$\leq \lambda u_1(x) + (1-\lambda) u_2(x)$$

which establishes the convexity of T.

Theorem :

For a random tree that is stationary in time, let $u \geq 0$ be a function in T as defined in property 6, and let $w(x) \geq 0$ be any function satisfying

(4.62) $$w(x) \leq u(x) - \mu_1(x, u(\cdot)).$$

Then, for all n

(4.63) $$\gamma_n(x, w(\cdot)) \leq u(x)$$

Proof :

Since $u \geq 0$, $\mu_1(x, u(\cdot)) \geq 0$ from (4.57), and thus

(4.64) $$w(x) \leq u(x)$$

Thus

$$\gamma_1(x, w(\cdot)) = w(x) + \mu_1(x, w(\cdot))$$

$$\leq w(x) + \mu_1(x, u(\cdot)) \leq u(x).$$

The first inequality follows from (4. 64) and (4. 54), and the second from (4. 62). We now use induction and assume $\gamma_n(x, w(\cdot)) \leq u(x)$. Then, from (4. 51)

$$\gamma_{n+1}(x, w(\cdot)) = w(x) + \mu_1(x, \gamma_n(\cdot, w(\cdot)))$$
$$\leq w(x) + \mu_1(x, u(\cdot)) \leq u(x)$$

This completes the proof.

Finally, if we look at the set T^* of functions w which satisfy (4. 62) for some $u \geq 0$ in T, we see that this set is convex also. For if

$$w_1(x) \leq u_1(x) - \mu_1(x, u_1(\cdot))$$

$$w_2(x) \leq u_2(x) - \mu_1(x, u_2(\cdot)),$$

then for $0 < \lambda < 1$

$$\lambda w_1(x) + (1-\lambda) w_2(x) \leq \lambda u_1(x) + (1-\lambda) u_2(x) - [\lambda \mu_1(x, u_1(\cdot)) +$$
$$+ (1-\lambda)\mu_1(x, u_2(\cdot))] \leq u(x) - \mu_1(x, u(\cdot)); \quad u = \lambda u_1 + (1-\lambda) u_2 .$$

The set T^* takes on greater significance when we look at the functions $w(x)$ such that $\lim_{n\to\infty} \gamma_n(x, w(\cdot))$ exists. If we associate $u(x)$ with this limit, then (4. 62) is satisfied with equality. In other words, T^* includes all functions for which this limit exists.

In order to make these results tractable, we must assume a special case now of a completely stationary random tree with freezing barriers at α and $-\beta$. We assume that each of the increments z_k for the particles branching from a given particle are statistically independent, and that the number of such particles is fixed at M_1. For the function $u(x)$ we choose

$$(4.65) \qquad u(x) = \begin{cases} \ln\left[1 + \dfrac{\delta e^{sx}}{1-\epsilon}\right] ; & x \in [-\beta, \alpha] \\ 0 ; & x \notin [-\beta, \alpha] \end{cases}$$

where s, δ, and ϵ will be determined later. Then

$$\mu_1(x, u(\cdot)) = \ln E\left\{\exp \sum_{k=1}^{M_1} \ln\left[1 + \frac{\delta}{1-\epsilon} e^{s(x+z_k)}\right]\right\}$$

$$= \ln \left\{\prod_{k=1}^{M_1} E\left[1 + \frac{\delta}{1-\epsilon} e^{s(x+z_k)}\right]\right\}$$

$$(4.66) \qquad = \ln \left\{1 + \frac{\delta}{1-\epsilon} e^{sx} E\left(e^{sz}\right)\right\}^{M_1}$$

$$(4.67) \qquad \leq \frac{\delta}{1-\epsilon} e^{sx} M_1 E\left(e^{sz}\right).$$

Observe that for the case we are considering,

$$M_1 \; E\left(e^{sz}\right) = E\left[f_1(0, s, 0)\right]$$

where f_1 is defined in (4. 7). In that previous discussion, we found that the process dies out with probability 1 if $r_{max} > 0$, which occured if

$$(4.68) \qquad E\left[f_1(0, s, 0)\right] < 1$$

for some range of s. We make that assumption here, and for some s satisfying (4.68), we select $\epsilon > 0$ by

$$1-2\epsilon = M_1 \, E(e^{sz}) \tag{4.69}$$

$$\mu_1(x, u(\cdot)) \leq \frac{\delta}{1-\epsilon} e^{sx}(1-2\epsilon) \, ; \quad x \in [-\beta, \alpha] . \tag{4.70}$$

Next we note from looking at the power series expansions that

$$e^y \leq 1 + \frac{y}{1-y/2} \leq 1 + \frac{y}{1-\epsilon} \quad \text{for } y \leq 2\epsilon .$$

Setting $y = \delta e^{sx}$ and comparing with (4.65), we see that

$$u(x) \geq \delta e^{sx} \; ; \quad \text{for } \delta e^{sx} \leq 2\epsilon . \tag{4.71}$$

It follows that

$$\mu_1(x, u(\cdot)) \leq u(x)\left(1 - \frac{\epsilon}{1-\epsilon}\right); \quad \delta e^{sx} \leq 2\epsilon .$$

This is satisfied for all $x \in [-\beta, \alpha]$ if we choose

$$\delta = e^{-s\alpha}\left[1 - M_1 \, E(e^{sz})\right] \tag{4.72}$$

Finally, we choose

$$w(x) = \delta e^{sx} \frac{\epsilon}{1-\epsilon} \tag{4.73}$$

and from the theorem

(4.74) $\gamma_n(x, w(\cdot)) \le u(x)$; all n, $x \in (-\beta, \alpha)$

This result can now be used to get an upper bound on the distribution function of the total number of progeny in the random tree. Let N_n be the total number of particles in the random tree out to time n, counting frozen particles just once. Assume that the tree starts at level 0 so that $x = 0$. Then

$$e^{u(0)} \ge \exp \gamma_n(0, w(\cdot)) = E\left\{\exp \sum_{\ell=0}^{n} \sum_{k} w \; y_{k,j}(\ell,0,0)\right\}$$

$$\ge E\left\{\exp N_n w(-\beta-\delta^-)\right\}$$

$$\ge \Pr\left[N_n \ge m\right] \exp\left\{m\, w(-\beta-\delta^-)\right\}$$

(4.75) $$\Pr\left[N_n \ge m\right] \le \exp\left\{u(0) - m \frac{2\epsilon^2}{1-\epsilon} e^{-s(\alpha+\beta+\delta^-)}\right\}.$$

Since this bound is valid for all n, it bounds the total progeny of a random tree with freezing barriers, and the bound is exponential in the number of particles.

It appears at first glance that the dependence of the bound on the upper threshold is specious, but some reflection shows that such is not the case. Any particle which gets close to the α threshold generates a cascade of particles which move down toward the β threshold, and thus it is not surprising that the tail of the distribution of N_n depends primarily on $\alpha + \beta$ rather than just β.

Contents

page

Preface 3

Introduction 5

Measures of Information 9

Finite State Channels 22

Source Coding with a Distortion Measure 50

Random Trees and Tree Codes. 84

Random Walk Techniques. 90

Branching Process Techniques. 101

Contents 115

Printed in the United States
By Bookmasters